Reshaping
Technical Communication

Reshaping Technical Communication

New Directions and Challenges for the 21st Century

Edited by

BARBARA MIREL
University of Michigan
Mirel Consulting

RACHEL SPILKA
University of Wisconsin–Milwaukee

Routledge
Taylor & Francis Group

NEW YORK AND LONDON

First published 2002 by Lawrence Erlbaum Associates, Inc., Publishers

Published 2015 by Routledge
711 Third Avenue, New York, NY 10017
2 Park Square, Milton Park, Abingdon, Oxfordshire OX14 4RN

First issued in paperback 2015

Routledge is an imprint of the Taylor and Francis Group, an informa business

Cover design by Kathryn Houghtaling Lacey

Library of Congress Cataloging-in-Publication Data

Reshaping technical communication : new directions and challenges
for the 21st century / edited by Barbara Mirel, Rachel Spilka
    p.   cm.
Includes bibliographical references and indexes.
ISBN 0–8058–3517–2
1. Communication of technical information. I. Mirel, Barbara.
II. Spilka, Rachel, 1953–
T10.5 .R48 2002
601'.4 — dc21   2001054528

ISBN13: 978-1-138-86128-2 (pbk)
ISBN13: 978-0-8058-3517-5 (hbk)

Contents

Foreword

JANICE REDISH

AN INSIGHTFUL BOOK FOR AN EXCITING TIME

This is an exciting time for technical communicators. As technology has caused major changes in most people's work, home, and play, the need for successful, professional technical communication has grown enormously. In 1991, the Society for Technical Communication (STC) had 13,778 members in 24 countries. In 2001, it had 21,789 members in 48 countries. Academic programs in technical communication have also expanded tremendously in the past decade.

The start of a new century is an excellent time to take a look at where we are and where we are going in both academia and industry. This book does that. With 11 essays and instructive introductions by the editors, *Reshaping Technical Communication: New Directions and Challenges for the 21st Century* provides us with insights on many aspects of the past, present, and future relationship between academia and industry and sparks a very interesting discussion on future trends for technical communicators.

COMMUNITY AS A THEME OF THE NEW CENTURY

If we look at these essays in a slightly different way, we also see that they are about *communities*. Community is going to be a major theme of this new century.

Many people in technical communication complain of feeling isolated— teaching and researching technical communication in a department that does not value it, working as a lone writer in a group of developers or even as the only writer in an entire company, working as an independent consultant in the isolation of a one-person office, or telecommuting and therefore working alone much of the time.

In some ways, the new technologies and the opportunities they give us are isolating: to work long distance, to move frequently, to study by oneself in an online course. And yet, in other ways, these same new technologies are helping to forge new communities—online communities as well as in-person communities.

A hundred years ago, for most people, the primary community was geographic. Family as community, profession as community, religion as community —all usually came together within a physical geography. Not anymore. Today, geography (neighborhood as community) is only one of many disparate communities that most of us belong to. And for many of us, our professional communities are primary.

This book highlights some of those communities:

- Communities of teaching.
- Communities of practice.
- Communities of research.
- Communities of users.
- Communities that bring different communities together.

Reality for all of us is that we belong to many communities—and, I believe strongly, this new century is going to broaden rather than contract many of these communities and is going to find most of us expanding the number of different communities to which we belong and with which we interact.

Communities of Teaching

A major theme of this new century is going to be *interdisciplinary* scholarship and teaching. Technical communication teachers already realize that their students need more than a background in rhetoric and a deep understanding of the writing process. They realize that technical communication is also about understanding users (from cognitive psychology), usability (from ethnography, anthropology, and human-computer interaction), and information architecture (from information science), as well as understanding information about information design and graphic design, technology, and so on. In the 1980s, many technical communicators clearly separated themselves from marketing communicators; in the new century, as many technical communicators are creating web sites, an understanding of marketing and branding has become essential.

While no teacher of technical communication is going to be interested in or expert in all of these different areas, the academic community of teachers of technical communication encompasses all these various and overlapping communities. Newer departments in most academic settings are, in fact, often set up by pulling people together from a variety of disciplines. And in many places, students are allowed to create their own majors by creating their own interdisciplinary complement of teachers.

The excellent essays in the first half of this book explain how different are the goals, rewards, and time between academic and industry communities and how we might bring these communities closer together. (See the contributions of R. Stanley Dicks, Ann Blakeslee, and Steve Bernhardt.) Within universities, communities of teaching across disciplines also differ.

For example, teachers of technical communication pioneered project-oriented courses and a process orientation that helps students develop generalizable skills in the context of specific products. Other disciplines—especially computer science—need to learn this. One of the major sources of frustration for many technical communicators in industry is that developers with backgrounds in computer science often have not had any coursework that prepares them for the team approach of the workplace, for working with and appreciating others' skills, for valuing communication, and for focusing on users and usability as well as functionality. Technical communication courses can serve as excellent models for other disciplines such as computer science. In the new century, technical communication teachers need to move beyond the confines of their own community to influence pedagogy in related academic communities.

Communities of Practice

Just as disciplinary boundaries within academia are loosening, and in some cases should be loosened further in this new century, boundaries of professions or fields within practice are also loosening.

As Rachel Spilka explains in this book, it has proven very difficult to define what a technical communicator does. That does not bother Spilka or me. People define themselves by the communities that they choose to join. And the plural *communities* here is critical. We are all members of many different communities.

I am a linguist by training and, therefore, know that we each belong to many communities of speech. Linguists call these "registers"—we speak differently in our office persona, in our home persona, in our parental (or child) persona, and in our "hanging out with friends" persona.

In the same way, both academics and industry people in any field or profession belong to many different communities: They are part of a departmental community—and probably part of a specific community within that department. They are part of a community of a specific institution or corporation (and we know how different the cultures can be from one university to another, from one corporation to another). They are part of an even larger community of academics from many different universities or of practitioners from many different corporations.

We can see our range of communities expanding outwards from ourselves as individuals in ever wider circles. We can also look at that picture in reverse and see ourselves in communities inside of communities.

For example, in 1991, STC had four Special Interest Groups. Today, there are 20 with memberships ranging from 97 to 3,714. These are communities within communities. They help people find like souls within what might otherwise be an overwhelmingly large community. But they also help people stay within the technical communication community who might otherwise have moved away as they changed jobs or skill sets. These special interest groups have kept our overlapping communities together.

In the second part of this book, Lori Anschuetz and Stephanie Rosenbaum talk about expanding roles for technical communicators. With each new role, a person engages in new and overlapping professional communities. We should encourage these overlapping communities.

Am I not a technical communicator anymore because I spend more time doing usability testing than writing manuals? As I see it, I am still helping people make products communicate. And although I don't write manuals anymore, I now help people turn old paper documents into usable writing on the web. And, I still teach lawyers how to rethink legal documents into clear communications, which I have been doing for more than 20 years. Yes, I've expanded what I do; I've moved into other communities. But I'm also still part of the technical communication community—and that community has itself expanded to include people coming with relevant skills from other fields.

I am a member of five professional organizations (the ACM Special Interest Group on Computer–Human Interaction [SIGCHI], the ACM SIGDOC, the IEEE Professional Communication Society, STC, and the Usability Professionals' Association)—and I probably should be a member of the professional societies of information architects, trainers, web developers, and writers of legal documents because I work in all those fields. How many communities do you belong to? My guess is more than one.

Isolation breeds sterility. Overlapping and intersecting communities bring new ideas. As this anthology argues, let us not worry too much about setting boundaries on who can be called a technical communicator. Let us instead welcome people who bring other community connections as they become part of this community.

Communities of Research

In technical communication, we have strong communities among academics and strong communities among industry people. One of our goals for the new century should be to strengthen our communities of research—within academia, within industry, and across those communities. As Deborah Bosley writes in this book, research informs practice and practice informs research and both can be done by faculty and students within the college or university setting. If you are in academia, read her examples and find similar research projects for yourself and for your students. If you are in industry, see how to take lessons learned from practical projects and apply their research implications to other projects.

Not only in technical communication, not only in English departments, but throughout the university, the "ivory tower" perspective is giving way to a new understanding that people who move beyond the university need to see the relevance to contemporary work of research and foundational theories and principles. We need to also work on the reverse — to have practitioners in industry understand the importance of basing their work on principles drawn from research. Many now do; at STC conferences, sessions explaining research in terms relevant to practice are usually jam packed with people wanting solid evidence and good advice. We also need to help industry practitioners realize the research potential of many of their own projects.

As Karen Schriver points out in this book, we have many stakeholders for whom research is essential. Schriver urges us to expand our sense of the communities for whom we are doing research and to whom we should be communicating about our research. She suggests that by doing relevant research and disseminating our research findings broadly and in language that the general public can understand, we could dramatically increase the awareness of the value of good information design. And that, in turn, could increase respect for what we do and funding for doing it well.

Communities of Users

Communities as a theme may also help us better understand the users for whom we write or for whom we help others develop products. Technical communication teaching and practice have always included an emphasis on audience analysis. We may speak of "the reader" or "the user," but we all know that documents have multiple audiences and products have many different users. And these users belong to various and overlapping communities. The notion of community helps us see the importance of the contexts in which those users live and work.

In the second part of this book, on broadening the vision, responsibilities, and influence of technical communication, you will find Barbara Mirel's essay on the importance of understanding users' work in context. The nurses of whom Mirel writes have software that allows them to do an efficient and effective job of a particular task — when all goes perfectly. In this case, that means that a particular patient is able to take all of the prescribed medicine in full at each medication time. But, in fact, the real world seldom works perfectly. What happens when the patient cannot swallow everything, when some medication spills, when the patient did not get a complete dose earlier in the day, when the nurse going off shift must leave a note about medication irregularities for the nurse coming onto the next shift? Then the new software for administering medications not only does not allow an efficient and effective job, it complicates the task. The software developers did not realize the importance of the actual work in context or of the communities in which the nurses work — how important communicating to each other about their patients is to members of this nursing community.

Similarly, Anthony Paré, in this book, tells us how he was not successful in training and working with Inuit social workers until he let them own the problems and the solutions in ways that worked for their community and their contexts.

The notion of context is critical. The European standard for usability, ISO 9241-11, for example, defines usability as "The extent to which a product can be used by specified users to achieve specified goals with effectiveness, efficiency and satisfaction in a specified context of use." As JoAnn Hackos and I wrote in our book on user and task analysis (*User and Task Analysis for Interface Design,* New York: John Wiley and Sons, 1998), we must understand the physical, social, and cultural contexts in which users live and work.

Furthermore, we need to understand that we have based most of our work on helping people use software, hardware, and web sites on incorrect assumptions about the contexts and communities of most users. We have built technical communication for products on a model of the lone reader. That model may be appropriate for novels and essays. You are probably reading this book by yourself. But carrying that notion of the reader in isolation into our work on helping products communicate is not appropriate. In fact, as we all know, most users needing help turn first to another person. They seek help within a community of people—not from users' guides or online documentation. We also know that in most communities, someone is likely to be known as the "guru" for a particular type of information. We may not yet know how to take the reality of the communal nature of learning and working into account as we build products and develop the communication within and for that product, but we should take this on as a challenge for the new century.

Communities That Bring Different Communities Together

This book contributes in major ways to our thinking not only about specific communities, but also about bringing communities together. Several of the essays are about sharing and improving dialogue between technical communicators in academia and in industry. We see this theme in essays like Paré's about interactions among the teaching community, the professional community of social workers, and the ethnic community of the Inuit. We can also see it in many of the essays in the second part of the book. Faber and Johnson-Eilola, for example, bring us a vision for the new century of university and industry communities working together as knowledge producers rather than as product communicators.

In the final essay in this book, Russell Borland—using a wonderful storytelling technique—urges technical communicators in both academia and industry to learn about interaction design, to understand contemporary technologies and tools, to learn the principles, practices, and nuances of the knowledge domains about which they write. In essence, he is inviting us to join these other communities.

Technical communicators have often served as the integrators of many different communities. We serve as advocates of a user community in the many cases when users have not been invited to the table. We bring together the subject matter experts with their domain knowledge, the developers with their tool knowledge, and the designers with their aesthetic knowledge and help all these communities see that the goal is to make products communicate successfully to users.

Perhaps it is our focus on communication that makes us see the value of sharing our knowledge and expertise. For example, in doing usability testing, I try always to work closely as a team with the product owners, planning the test with them, having them as observers and note-takers when I am conducting the test, working jointly with them in analyzing the data and deciding what to do to improve the product. One of my goals is to have them learn how to do usability testing, to want to do it, and to gain skills in doing it well.

Other usability specialists and many people in other professions don't believe in this team approach. They feel that they must hoard their knowledge to be seen as experts. If everyone knows how to do what they do, what will their expertise be?

I don't see it that way. To me, sharing one's knowledge and expertise does not diminish one's professionalism nor one's claim to a place on the team. In my experience, mentoring others to write well, to build usability into products, to focus on communication has many benefits. It makes products better. It reduces rework and the need to hurt others' egos by telling them they did it wrong. It strengthens inter-personal and inter-community relationships. And it increases the appreciation that people in other fields have for what we do. It also brings us more work—because they see the value of good communication or usability studies, want more of it, don't have the time to do it all, and realize that even though they have learned some, we still are more expert at it. Expanding our communities to invite others in only helps both them and us.

ENJOY THIS BOOK

There is much to think about in these 11 essays and in the editors' introductions. The editors tell you that these 11 essays do not cover all the issues. That's fine. The book is meant to make you think in new ways and to inspire you to continue the discussion about the issues raised here. May it also inspire you to raise other issues and to perhaps contribute to another volume in a few years as we see how this new century progresses.

Preface

We began this project with the goal of inspiring change. In our continued attempts to meet this challenge, we devoted up to 2 years collecting and then helping authors develop the chapters of this volume.

The project began in April 1999 when we issued a call for abstracts for the anthology at the concurrent Annual Conference of the Association for Teachers of Technical Writing (ATTW) and Conference on College Composition and Communication (CCCC) in Atlanta. Soon afterward, we issued the same call for abstracts on three Internet electronic mailing lists, those of ATTW, the Association of Business Communication (bizcom), and a usability testing group. Although we received many outstanding abstracts, many of them focusing on classroom projects and other relatively small-scale successes in the field, we decided early on to focus the anthology, instead, on larger scale, nontraditional ideas for moving the field forward in new directions. We also worked hard to achieve a reasonable balance between chapters written by academic and industry specialists in the field.

To move the anthology closer to our vision of what it might achieve, we decided to invite technical communication specialists from both academia and industry who we considered especially creative and innovative in their thinking to contribute new chapters. We also coordinated and hosted a June 2000 Milwaukee Symposium of 18 such specialists (see the end of this Preface for a list of these attendees) to discuss the current status and future of the field. We hoped that from this gathering would emerge several more chapters for the anthology and new insights for authors already planning chapters for the collection. As a result of all these efforts, we were successful in assembling the chapters in this volume. The Symposium efforts also led to the material included in the Appendix. There, we chart the research questions and issues that Symposium participants deem crucial for the field to investigate in order to make greater progress in the coming decades. At the Symposium, industry and academic specialists alike valued these research areas, and they hoped that investigations would be carried out by specialists in both worlds, sometimes separately, sometimes in concert.

We are excited that most of our contributors have moved quite a bit in their careers between academia and industry and have brought to their chapters special insights about both worlds. Throughout the experience of developing and editing this collection, we have also found it extremely useful that the two of us have had considerable experience in both academia and industry. Our hybrid backgrounds have helped us appreciate the special dimensions, needs, and potential of both worlds. We are proud that our contributors include some of the best thinkers and most exciting leaders of the field. We predict that readers will join us in being greatly impressed by their visions and insights. We do realize that quite a few more of our contributors have experience in the computer industry than in other realms of technical communication (e.g., government, business, health, environmental, or social service writing), but we have tried to represent other kinds of technical communication, as well, throughout the anthology.

Of course, reform involves conflict and frustration as one of the basic components of team-driven innovations. We encountered many differences of opinion at the Milwaukee Symposium and even in our own collaboration on this project. We often discovered that we held differing points of view about what might interest different types of readers from academia and industry, about how to make descriptions meaningful for all constituents in our multiple audience, and about how to shape the volume so that it would deliver a vision that all our readers would consider unified and coherent. We will leave it to our readers to judge how well we resolved these differences. We would, however, like the chance to thank each other for remaining honest, respectful, and patient during trying times and for our enduring friendship throughout.

ACKNOWLEDGMENTS

We are grateful to Linda Bathgate at Erlbaum for her generosity, graciousness, patience, and support throughout this effort. Her enthusiasm for the innovative nature of this project has meant a great deal to both of us, as has her consistent belief that this anthology could help move the field forward in ambitious, important ways. We are also grateful to the following participants of the June 2000 Milwaukee Symposium for their exciting insights, which led to important new chapters for this anthology and contributed greatly, as well, to our own thinking for this anthology: Stephen Bernhardt, Russell Borland, Deborah Bosley, Stan Dicks, Roger Grice, Kathy Harmamundanis, Johndan Johnson-Eilola, Susan Jones, Jimmie Killingsworth, Leslie Olsen, Jim Palmer, Judy

Ramey, Mary Beth Raven, Stephanie Rosenbaum, Karen Schriver, and Elizabeth Tebeaux.

Finally, because this project was pervasive in our lives over several years, we are indebted to the encouragement that we have received from mentors and to the great love and support that we have received from our families. In moving into industry after years in an academic career, Barbara is extremely fortunate to have found an outstanding mentor in her manager and friend, Paul Schuster. His enthusiasm, vision, and commitment to improving products and processes have inspired her ever since. In her work on this volume, she thanks her husband Jeff, deeply treasuring the love and companionship that help them to share all of their projects. Jeff's endless support, thoughtful editing, and exceptional listening, advice, and ideas enriched this volume. Rachel is grateful to her father, Mark Spilka (1925–2001), for modeling for her not just the courage but also the grace and humility needed to take on controversial projects like this one. He was by far the most extraordinary teacher and inspiration of her life. She is also extremely proud of her children, Rahel and Zahara, for learning to play so well independently (and sometimes, miraculously, together) while she worked on this project at her home computer. Their unconditional love kept her strong and happy throughout this project.

Introduction

In the 20th century, the field of technical communication made tremendous strides. What began as a specialty matured into a field. In many organizations, what began as one or a few writing or editing positions became entire departments or divisions of documentation, online help, or technical communication. In many colleges and universities, what began as just a sprinkling of academic courses became entire technical communication programs or even departments. At the Society for Technical Communication, what began as a modest assembly of technical communication specialists grew into an impressive, powerful organization recently boasting more than 24,000 members. In the latter half of the 20th century, the field strengthened to the point of heralding its own body of empirical studies on technical communication and related topics, its own journals and anthologies, and its own annual regional, national, and international conferences. By the end of the century, it was no wonder that many technical communication specialists were proud and excited about how far the field had evolved over the past fifty plus years.

The field, however, cannot afford to rest on its laurels. The 21st century challenges us with unprecedented demands. Digital information and electronic records have become ubiquitous and have given rise to new genres, new media beyond print, new modalities beyond words, and new expectations for quick turnarounds. More and more frequently, content and structure are divided between content strategists and information architects and integrated by yet other information designers, often threatening the cohesion, coherence, and unity that make communications effective. Modules and re-use have become watchwords in the composing of communications, making it difficult to tailor compositions to specific audiences who, from a marketing point of view, now require more personalization than ever before.

On a larger scale, information systems and electronic records have proliferated at a pace that has outstripped organizational abilities to integrate them across functions. Yet technical communicators need to help audiences understand the uses, meanings, and implications of cross-functional information. In addition, as boundaries grow fuzzier each year between technologies and communications about them, technical communicators now share their claim to audience, task analysis, information design, and evaluation with other colleagues, for example, those in branding, interaction design, systems engineering, product management, and graphic arts.

All the while, despite acclamations of progress, communications in technological settings and about technological products have not necessarily progressed in quality, meaning, or effectiveness. Exchanges of knowledge and decisions based on them have not risen to new heights. Audiences have not smoothly —if at all—assimilated their technologies into the commonsense knowledge and social practices that give value and meaning to their work. At the heart of technical communication lies the vision of assuring effective and inclusive communications sensitive to audiences' social and cultural contexts. This vision is as pressing as ever and still needs to be realized. But given the complications of today's worlds and workspaces, technical communicators must redefine what this vision means in relation to current trends and contexts. To do so, we must reshape our roles and contributions so that we take a lead in identifying and designing the improvements that are still in dire need of progress.

As we look back at the past few decades, we believe that the changes that will be needed for technical communication to survive and prosper in the 21st century are profound. In recent decades, along with its accomplishments, the field has been experiencing considerable difficulty keeping up with the times. Progress in our field's research, theory, practice, and pedagogy—as important as it has been—has not kept pace with either the transformations wrought by the technologies with which we work or the growing demands for effective, valuable, and satisfying interactions with technologies and information systems. We have no choice now but to reinvigorate the field.

This anthology aims to assist this reinvigoration. It points toward new directions for greater growth and influence of the field. This collection emphasizes that the field must reassess and revise its status, identity, and value in order to strengthen and reposition itself and fare well in the 21st century.

Collectively, the chapters in this anthology explore two central questions for our future:

- How can we strengthen internal relationships between technical communicators in industry[1] and academia and external relationships with cross-functional colleagues in ways that help elevate our status and value?

---

[1] For the sake of simplicity, we are using "industry" throughout the anthology to refer to any professional and non-educational contexts where technical communication does or could take

- How can we broaden our responsibilities and influence to become major players within and across work contexts, thereby extending ourselves beyond our traditional roles as designers and composers of communication products?

We have organized this anthology into sections that respectively address these two questions. Part One focuses on internal challenges and evolutions in the relationship between academia and industry. Part Two casts its gaze externally—to relationships between technical communicators and other stakeholders—and suggests new approaches for assuming expanded responsibilities and leadership roles in our organizations.

Part One chapters describe and analyze the relationship between academic and industry specialists in technical communication. Improving this relationship will be critical for the future of our field. These two worlds will need to be in sync for the field as a whole in order to achieve greater value and status.

Unfortunately, our field remains quite distant from that goal. Stereotypes about academics and practitioners abound in our field. Many industry professionals, for their part, have claimed for years and still claim that academic research does not meet their needs because it is overly theoretical, focuses on the wrong issues, and is years behind the times in regard to the effects of tools and technology on information development and knowledge management. Academics have long portrayed practitioners as primarily concerned about how-to techniques with little regard for conceptual frameworks and as proprietary about their work and industry advances. Why are these two worlds traveling in parallel with so few connections, especially when they share the goal of developing a significant voice in the workplace? More importantly, how might academics and practitioners work together, instead of separately, toward the advancement of the field?

These questions are at the heart of what Part One contributors address. In analyzing our current distance from achieving the goal of creating strong academic-industry relationships, the contributors focus on the complexities of these relationships and on differences and similarities between the two worlds. They argue that to effect significant change in the field and unite toward common goals, academics and practitioners will need to overcome some serious barriers.

Yet, problems that technical communication specialists face in regard to insufficient influence and overly constrained contributions cannot be resolved simply by improving dialogues within the discipline across the two worlds. Instead, to change our status profoundly, we will need to modify ways in which we situate ourselves as influential agents both within our respective institutions and within and across our cross-disciplinary communities. Technical communi-

---

place. These contexts could be in government, health, business, manufacturing, computer software companies, or any non-academic work setting.

cators, after all, do not exist in a vacuum: other people also have a stake in our projects and areas of interest, and they have been competing for and gaining authority over processes central to our work such as writing, usability testing, quality control, and software design and development. Technical communication is just one of many professions now seeking to assume primary roles in interaction design, information development, and knowledge management. Intentionally or not, organizational structures of work often militate against the genuine collaborations that are necessary to advantageously bring together the unique contributions of all of these professions toward a common objective. How will the tensions between control and collaboration be negotiated, how will relations of power be distributed, and what will be the position and fate of technical communication?

Clearly, technical communication is experiencing an identity crisis. The unique strength that technical communication specialists bring to their projects is that they put a rhetorical stamp on the dramatically changing technology of workplace communication. Within projects, this process of shaping technologies and communications to the demands of context, purpose, audience, and medium may be called interaction design, contextual inquiry, information architecture, content strategy, information development, usability, or knowledge management. The challenge for the coming decades is to show our workmates the unique knowledge and skills that we as technical communicators bring to these areas and to assume roles of leadership.

Part Two takes up this theme as it explores external relationships between technical communicators and their colleagues, managers, and project stakeholders. The main message of Part Two is the need for technical communicators to move from the traditional circumscribed status of document writers and editors to more elevated and expanded positions in which they are vital to the strategic workings of their organizations and to the directions and designs of their technological products. Some individual instances of our strategic influence and knowledge creation exist at present. But they will have to become common to the field at large. As a whole, the field must become associated with strategic planning and decision-making that reaches beyond publication departments into product management, product design and development, and cross-disciplinary research projects. In addition, to assure that products as a whole and not just the documentation or interfaces embody support for users' and readers' needs and practices in their contexts of work, the field will need to influence the processes of production. Finally, technical communication specialists must assume a principle role in conducting research on user experiences and reader responses, setting design requirements, and in evaluating the usefulness, usability, learnability, and even enjoyment of technological products and documents.

Evolving the field in these ways supplants traditional definitions of what we do with a more expansive vision of what we could and should become. Gaining recognition and status for the value of our contributions is both an external and internal enterprise. Internally, as many chapters reveal, technical communicators in both worlds have to recognize and build on strengths that they often view as shortcomings, such as diversity. Externally, technical communicators cannot expect that a new professional role and vision will come about by carrying out business as usual, with "reform" efforts taking old forms, even if they are ramped up versions. Instead, changing external perceptions will require concerted efforts, for example, in how and to whom we spread "the word" about what we do, in how we (re)structure organizations to expressly promote this vision, and in how we (re)position ourselves strategically within development processes and as boundary spanners in the creation and exchange of mission-critical information and knowledge.

In describing the field's current status, in re-articulating goals for the field, and in exploring more expansive ways of achieving those goals, Parts One and Two together propose ways to re-define the field's very identity. They propose new parameters for identifying who we are and why we exist, what we should research, how we should do our work, and how we should disseminate our empirical findings and unique expertise. In brief, they propose reshaping and thereby reinvigorating the field.

We recognize that in looking forward, the contributors have not yet addressed some important issues that will be crucial to our professional growth and success. These issues include:

- Political strategies within our workplaces to increase our value, influence, and authority.
- The need for funding in order to support new directions in research and practice and the politics of current funding sources and criteria.
- Strategies for putting together and articulating jointly shared goals and objectives and jointly shared agendas for investigating pragmatic problems critical to the field.
- Skills in moving through the processes of forming-storming-norming-and performing in working jointly on innovations and being able to reap creativity from a healthy dissonance among perspectives.

The central purpose of this anthology is to propose far-reaching, innovative, and nontraditional strategies, visions, and ways of thinking. The anthology authors do not offer concrete solutions, recognizing that it is premature to do so before large-scale, profession-wide dialogues and efforts are launched. Because some of the authors' arguments and proposals challenge steadfast assumptions in the field, we expect the anthology to generate some controversy. Readers may respond in various ways, ranging from denial to skepticism to enthusiasm. We welcome all reactions. Above all, we hope that our anthology will inspire readers to continue the discussions and debates introduced here. We realize that many, perhaps most of our readers are invested heavily in the future of technical communication. Perhaps this anthology can help them find ways to contribute, personally, to reinvigorating and reshaping this field to ensure its continued vitality into the new century.

REVISING INDUSTRY AND ACADEMIA:
CULTURES AND RELATIONSHIPS

As we look forward to the next few decades, the relationship between academic and industry specialists is just one of many ingredients of growth and development in technical communication. However, it is an important one because professionals from both worlds contribute to the substance and identity that technical communicators hold dear within the field and in the outside world. Part I reveals explicit or subtle ways in which the two worlds can earn each other's respect and overcome cultural divisions in order to identify and accomplish shared goals.

In examining academic–industry relationships, the Part I contributors focus on complex differences and similarities. They argue that academic and industry specialists need to overcome serious barriers before agreeing on common goals and creating powerful allegiances. With this challenge in mind, the contributors explore such questions as these:

- What can be done to overcome cultural barriers between academic and industry worlds, and between the various worlds within industry?
- What might broad coalitions do to improve workplace practice from the technical communication perspective?
- How can academic and industry researchers ensure that empirical results from studies in both worlds will be relevant and of value to each world?
- How can academic and industry specialists reach a consensus perspective, and then strengthen their (united) voice in all stages of work life in workplace contexts?

Contrary to current threads of thought in the field, the Part I authors contend that professionals in academia and industry need to go beyond traditional solutions for easing tensions, improving dialogues, building bridges, and strengthening bonds between the two worlds. These solutions have included research projects, consulting, student and faculty internships, advisory boards, and service learning. In Part I, a strong theme is that strengthening the academic–industry relationship and elevating the status, influence, and value of the technical communication profession will require solutions that are more ambitious, broader in scope, and farther reaching. Toward that end, each contributor to Part I proposes conditions for accommodating more dynamic and flexible academic and industry contributions to training, research, and practice.

The first few Part I contributors expose prevalent stereotypes in the field about academics and practitioners. Stanley Dicks identifies primary cultural differences between the two worlds that have led to false, stereotypical impressions of each other. These differences, Dicks argues, have also had the unfortunate effect of discouraging or defeating many attempts at academic–practitioner collaboration. Dicks ties this absence or curtailment of collaboration to differences in the two worlds' perceptions of information, language and discourse styles, views of collaborative versus individual efforts, assumptions about employment, and reward structures. Dicks hopes that once technical communicators become aware of potential cultural differences, they can use that knowledge to prevent or overcome breakdowns in communication, understanding, and collaboration.

For Dicks, the most effective kind of academic–industry collaboration is one that "transgresses fewer of the cultural divides." For example, one reason that internships, usability testing, and industry visits to classrooms have been so successful is that these types of collaboration are short term, mutually beneficial, and do not challenge "either culture's basic principles." In contrast, lengthier projects that provide no short-term benefits and require more project management and communication are more challenging in terms of averting conflict. Dicks explains that for broad-scale types of academic-industry collaboration to succeed, it helps to define expectations and outcomes from the beginning, including which outcomes will be considered proprietary and which can be published, and to decide, right from the start, commitments of resources, people, time, and finances. By identifying, from the onset of joint projects, possible cultural impediments to successful collaboration, collaborators across worlds may develop strategies accordingly and increase the chances of a project's success.

In contrast to Dicks' emphasis on differences, the next two chapters focus on similarities and areas of overlap between academia and industry that offer benefits we often overlook. Deborah Bosley contends that dwelling on differences between academics and practitioners can thwart successful collaborations and partnerships between the worlds. From an academic point of view, Bosley worries that technical communication specialists in the university may hold traditional assumptions about differences that "in unproductive ways" keep them from disseminating their research to industry.

Bosley proposes that academics hoping to extend their influence beyond the university first identify common ground between academic and industry work environments, work practices, and writing habits and products. Then they need to make perceptual and behavioral changes in how they define themselves and do their work. For example, Bosley suggests that academics define themselves not just as teachers and researchers, but also as practitioners, as technical communicators who can work on documents and communication projects within both university and local communities. Doing so may increase the status and value of academics and help them earn the respect of practitioners. Bosley also urges a change in publication habits. She foreshadows Karen Schriver's chapter in Part II by proposing a more expansive dissemination of research findings. Instead of reporting research findings primarily in academic journals, academics need to consider practical applications of their findings and disseminate those in publications that are accessible to a practitioner audience. Bosley concludes, "It is *only* through this kind of recognition—that each community has something to offer the other—that technical communicators will truly respect each other and want to collaborate and partner together for life-long learning."

Continuing Bosley's emphasis on similarities between academia and industry, Ann Blakeslee makes a case for shifting the focus of our research to the "overlapping space" between academia and industry. As Blakeslee puts it, this "overlapping" or "boundary space" between the two domains "suggests an area in which language, rhetorical aims, and work processes might be held in common." In this space, dialogues between academics and practitioners may occur that generate richer and deeper understandings, leading to joint discoveries of shared goals and new means for mutually achieving them. Blakeslee urges a program of academic research to help the field identify, study, and understand the traits and workings of a "common ground" with this generative potential. She provides us with a look at what this kind of research might involve by describing projects that she conducted on classroom–industry collaborations at two universities. Blakeslee's post hoc analysis of these projects reveals new approaches that researchers who study industry–academic collaborations can take in order to uncover subtle yet crucial new ground—overlapping spaces—for mutual support and advancement. These new approaches involve highlighting and negotiating the social and political dimensions of the communications and "deliverables" that are exchanged among students, teachers, and industry sponsors.

From her two case studies, Blakeslee illustrates ways in which nuanced differences between each world's genres of project communications are likely to impede the two worlds from working together productively to create innovative products. To achieve project processes and dialogues that enhance joint goals and mutual support, Blakeslee recommends more research on classroom–industry projects directed toward discovering new knowledge of genres and toward mutually negotiating genres across worlds. What is gained from this kind of research is a richer appreciation of the subtleties and complexities of technical communication, including the social and political dynamics that play out and affect products.

In the next chapter, Anthony Paré illustrates how a participatory-action approach to research also has great promise in deepening the field's understanding of writing in particular social contexts. Paré describes how, after 15 years of teaching, training, research, and consulting work, he discovered that he could claim just modest success in influencing workplace practice. As he puts it, "Nothing changed as a result of my work." The reason was that his teaching or training was done at a distance from the social contexts that concerned his inquiry. Paré sought to "acknowledge the [social] embeddedness of workplace writing—indeed, to exploit it" by conducting participatory-action research, in which participants "set the research agenda, participate in data collection and analysis, and exercise control over the whole research process." Paré describes one such project in which he asked Inuit social workers in northern Canada to define the problems of their field. According to Paré, this type of research, by allowing the social workers to create their own power, led to a far more complex and rich understanding of workplace writing than the social workers or researchers could have otherwise achieved. As he puts it, participatory-action research "made it possible for all of us as a group to negotiate a space between cultures, a space where teaching and learning were the natural outcomes of a common and collective need to know, and there the roles of teacher and student were constantly interchanged. In the process, we were all transformed."

Just as Blakeslee and Paré propose new, more expansive types of research that aim, at least partly, to facilitate greater understanding between academics and practitioners, Stephen Bernhardt focuses his chapter on describing a more dynamic type of collaboration that has the potential to dramatically strengthen two-way bonds between academia and industry. Bernhardt echoes Dicks's sensitivity to cultural differences between academics and practitioners. He sees valid reasons for the two worlds to remain separate in many goals and practices. Instead of achieving a complete unification of the two worlds, Bernhardt advocates what he calls "shared communities of practice involving frequent, active, project-based cooperation." He argues that only by working together through project-based activities will academics and practitioners develop the knowledge and concern about each other necessary for successful collaborations toward shared goals. Bernhardt calls this *active-practice,* which he defines as the creation of productive tension between academia and industry. For example, while practitioners spend time on campus, teaching and working with students, faculty and students can spend time in workplace jobs. Together, practitioners and academics can combine their expertise as they collaborate on research projects. The result, according to Bernhardt, would be significant: "a tempering of distant, academic critical posturing and industry skepticism, together with a recovery of relevance and understanding across the divide."

One concern surfaces from all chapters in Part I combined: What is the most inspiring metaphor for referring to goals that the profession needs to pursue regarding academic–industry relationships? Should it be "building bridges," "narrowing gaps and differences," "meeting in overlapping spaces," or paradoxically, "finding unity in difference?" Contributors to Part I show that quick and easy answers to this issue do not exist, because academic–industry relationships are enormously complex.

Perhaps most centrally, Part I argues that whether the academic–industry relationship is a gap or not is less a concern than whether these two worlds can find ways to pull together toward the common goal of improving the field. According to Part I contributors, there is value in developing our knowledge of both differences and similarities between academia and industry, and then in using that knowledge to find more innovative, expansive ways to unite the two worlds. Doing so would help all technical communicators work together, as a unit, both toward internal goals such as improving research, theory, practice, and training, and toward external outreach efforts such as finding ways to increase their status, value, respect, and influence in workplace contexts. Consistent with a key purpose of the anthology as a whole, the Part I chapters aim to extend our thinking about the future of our field, and to develop a more expansive vision of how academics and industry specialists in our field might benefit from working together, instead of apart, to identify and then pursue goals that both worlds have in common.

1

Cultural Impediments to Understanding: Are They Surmountable?

R. STANLEY DICKS
North Carolina State University

Collaboration between academics and practitioners in technical communication is essential to both groups. Academic programs in technical communication came into existence because of the needs of business and government for competent communicators with the special knowledge and skill sets required to produce technical documents of high quality. Those programs, in a real sense, depend for their existence on continuing to meet the needs of the "work world." To ensure that their programs continue to meet such needs, academics must continue to communicate with and collaborate with practitioners. In turn, business badly needs to hire technical communicators trained in the special requirements of audience-centered writing as opposed to the journalistic and expository writing instruction received in more traditional writing programs. Corporate and government entities, which face a constant shortage of qualified, competent communicators, must rely on academic programs to help supply enough such practitioners. In addition, to improve the quality of their work, practicing technical communicators need to interact continuously with academics to remain informed of the results of academic inquiry and research into appropriate principles and practices (Tebeaux, 1996).

Although both groups benefit greatly when they interact and collaborate effectively, many cultural differences between academia and business thwart collaboration efforts. From 13 years in academia and 16 years as a practicing technical communicator in industry, I have seen how those differences can discourage or defeat what would otherwise be mutually beneficial collaborations. I have seen groups decide not to collaborate because they could not agree on whether to publish the results. I have seen potential collaborations break down because business management assumed that they would not get anything quickly and beneficial enough to warrant the time and expense.

To understand how to collaborate successfully, we must first understand how our serious cultural differences mitigate against it, and we must then work toward finding ways to circumvent those differences. They can be so pronounced that sometimes even communicating can be difficult. If we attempt to collaborate without being aware of these differences, we risk breakdowns in communication and understanding.

What are the most significant differences and how do they affect our attempts to work together? In this chapter I discuss cultural differences in five main areas:

1. Perception of information, including radically different perceptions of its value and dissemination.
2. Language and discourse styles.
3. Views of collaboration versus individual effort, including different collaborative models and reward systems for collaboration.
4. Assumptions about employment, including differences in perception of time, fealty to a discipline or an employer, the role of research and publishing, workloads, and the use of power.
5. Reward structures.

Given all these differences, it is possible for academicians and practitioners to communicate and collaborate, but in many cases it may require, at least initially, focusing on short-term arrangements rather than long-term attempts that tend to confront too many cultural differences. The chapter concludes with a discussion of short-term methods that can help avoid the differences and lead toward successful collaborations that are so important to both groups.

THE PERCEPTION OF INFORMATION

Academics perceive information as something to share. They are required to share information with students; their teaching evaluations are often based on how well they do so. Furthermore, they are required to share their research with colleagues in books, journal articles, and conference presentations. If they do not do enough of this sharing, they forfeit tenure or promotion. One of the very basic foundations of education is that information should belong to everyone and be shared by everyone. The idea that someone can own information is alien to many academicians.

Practitioners, on the other hand, soon learn that information has monetary value. Indeed, as we enter the information age, many companies increasingly view their only real asset as information they have that others do not. Most companies have guidelines, often stringent, for protecting their intellectual property, copyrights, and trademarks. As the Web increases its role as a repository of shared information, a new kind of information broker is emerging. Owning the means for getting to information (such as a Web portal) can be even more valuable than the information itself.

The business tenet that information is proprietary leads to another disjuncture with academia. Even those few people in the business community who might be predisposed to participate in academic dialogue are discouraged from doing so. In my 16 years in business, I was encouraged to attend professional conferences to find out what others were doing and saying. However, I was discouraged from presenting information about the results of organizational research and development, on the basis that those results were valuable to the organization and should not be shared outside of it. It was considered appropriate to give a general presentation that described the organization's activities and capabilities in the hope that it would generate customer leads and sales. But it definitely was considered inappropriate to share real research results externally. This perception about the nature and value of information, along with normal time pressures that most practitioners face, militates against practitioners participating in academic dialogue. Even so, practitioners need to know the results of academic research to keep current with the latest thinking in the discipline. For that reason, an entire business of training classes, consultants, and writers has emerged to act as intermediaries who study the academic literature and the more popular literature for trends in the discipline, and then translate the results into workshops, conference presentations, and books designed to be immediately useful for practitioners. Unfortunately, even though academics are capable of preparing such materials themselves, they rarely do so because it would not help them toward tenure and promotion.

Often, this fundamental difference in the perception of information and its value severely restricts collaborations between academia and business. Businesses worry that academicians will publish results of the collaboration that reveal proprietary information. Academicians cannot understand why businesses so jealously protect information that, if published and made available to everyone, would make the world a better place.

LANGUAGE AND DISCOURSE STYLES

This topic warrants its own book. Both communities have their own discourse sets, and each finds the other's to be laughably absurd. Business people make fun of academic obfuscation and political correctness; academics make fun of

TABLE 1.1
Business and Academia Discourse Differences

Business & Industry	Academia
Active	Passive
Concrete	Abstract
Inductive	Deductive
Instrumental	Rhetorical
Product, result	Process
Presentational	Reflective
Direct	Indirect

people in expensive suits saying "proactive" frequently. The language sets say much about the different cultures. Business lingo includes much terminology associated with finance and project management, such as budgets, head counts, directs and indirects, resources, deliverables, and milestones. This language set stresses speed and efficiency. Meanwhile, the academic lingo tends to include many abstract terms and concepts that have nothing to do with practical outcomes, such as rhetorical situations, cultural artifacts, writing heuristics, and discourse communities.

Members of an academic department are colleagues; in a business department, they are coworkers. Members of an operations group in academia are administrators; in business, they are managers. The difference between one who administers and one who manages says much about how power is viewed in the two worlds. Table 1.1 (subject to all the dangers of generalization) contrasts the natures of the two discourse communities.

As a result of these language differences, the two groups often talk right past each other when they try to collaborate. As always, language differences represent cultural differences. Hence, academicians may have trouble following business discussions about, for example, "critical path deliverables," laden with project management terminology. And people in business may not understand academic concerns about, for example, how documents must reflect various rhetorical considerations inherent in discourse communities.

VIEWS OF COLLABORATION VERSUS
INDIVIDUAL EFFORT

Nearly all practitioners are part of a team, group, or unit of some kind that works collaboratively (Barchilon & Kelley, 1995). At many companies, all work is collaborative, and rewards are based on the success of collaborative efforts. Academicians, in contrast, are rewarded with tenure, promotions, and plum teaching assignments based more on individual efforts. Although they are typically encouraged to work with others on committee assignments, tenure and promotion decisions almost always are made solely on the basis of individual research and publication efforts.

It is becoming common knowledge in business that the old, militaristic, male-dominated, hierarchical system of management does not work very well. In most companies, that hierarchy is being slowly dismantled and replaced with more collaborative structures. Perhaps academia was ahead of business here, as it has operated on a more collaborative model for a long time (although motives for doing so are due more to a distrust of power than a yearn for efficiency).

However, there is a difference in the types and goals of collaboration. In business, collaboration is done to get results and achieve goals as quickly as possible. In academia, it is more often done to delay results, to ensure that all possible contingencies have been studied before anyone acts. Whereas in business collaboration is results- and product-driven, in academia it is almost wholly process-driven, often without any specific result or goal in mind (". . . we've always had this committee").

When business and academia collaborate, complications grow. Nearly always, a management person decides if and how the business will collaborate. Managers are trained to get results, not to make long-term investments. Meanwhile, academics believe in investing in the future: They train people to think and analyze better and do research to contribute to a better future. When academics see that managers want immediate results, paybacks that are objective, empirical, and measurable in the short term, they become disappointed or discouraged. Similarly, business managers become frustrated when academics show little concern about producing definitive results quickly.

ASSUMPTIONS ABOUT EMPLOYMENT

Cultural differences between academia and business cover nearly all aspects of work and employment. First is the very assumption of employment. Whereas academics see themselves as *permanent* members of a discipline and officers of instruction, practitioners see themselves as working in positions that they will probably (and often intend to) leave someday, at a company that they may also leave someday. Many academics get as much (or more) positive reinforcement for their activities from their discipline as they do from their institution. Most practitioners get little or no reinforcement from their discipline; rather, it comes from their company, department, boss, or paycheck. Their fealty to those sources of reinforcement can undermine attempts at collaboration; they are apt to look for a rapid payoff for their organization rather than the chance to publish results for other members of their discipline.

PERCEPTION OF TIME

In business, management control and monitoring actively direct daily, weekly, monthly, and annual work patterns. In academia, once teaching assignments are made for the year, little supervision of one's activities takes place. One is expected to complete teaching assignments and scholarship successfully during the year, but monitoring of results is loose or nonexistent compared to business. Close perusal of results may occur only every few years. Then, once tenure is awarded, it may not occur at all unless the faculty member is applying for a promotion.

Perceptions of time also differ in terms of goals and objectives. Business activity is driven by short-term goals and objectives. In academia, semester-long teaching assignments are short-term goals, whereas tenure (6–7 years) and promotion to associate and full professor (measured in years or decades) are long-term goals. Daily work in academia, outside of teaching, is a result of love for (or, in some cases, slavish devotion to) the discipline and the goal of tenure or promotion. Daily activities in academia can be just as hectic as those in business, especially when they include advising, committee service, and administrative work. Such service is sometimes mandated, but at many institutions a faculty member can choose to do very little of it. These differences in time perspective can impede smooth collaboration between business and academia. Academics cannot comprehend how business people can insist on short-term, perhaps incomplete, solutions, when it is obvious to them that a problem needs to be studied longer. Similarly, business people cannot understand why academics cannot arrive more quickly at firm, decisive answers. I know of one manager who cut off a collaboration on multimedia on the Web because he was not getting any short-term return and feared that he would simply fund ongoing research for years without getting any concrete information or results to aid in developing viable, online, multimedia solutions.

WORKLOAD

Business people perceive that they work harder than academics. They tend to envy the month-long Christmas "vacations" and 3-month summer "vacations" that academics get, even though they understand that academics must devote those periods to program development, course design, class preparation, research, and publishing. Some tenured faculty teach their classes the same way each year and have stopped publishing, so that they indeed get 4 months a year "off." Many academics work every day of the year, including weekends, rarely or never taking vacations and putting in hours that few business people do. On the other hand, the work pressures that practitioners experience tend to be intense and relentless. In many companies, especially technical ones, practitioners work considerably more than 40 hours a week. My observation is that technical writers average more overtime than most groups because of the deadline-driven nature of the job. Hence, their workload leaves little or no time for contemplation, theorizing, or studying the issues and concepts behind their discipline. Very few practicing technical communicators have the time to read scholarship in the field and stay abreast of the latest research. The pressure on their time is further exacerbated by rapid changes in the media they publish in and the tools they use to do their work. They must continually learn new technologies and software programs and can rarely afford the time it takes to study what academics in the field are publishing (Kunz, 1995). Not surprisingly, collaborative conversations, interactions, and projects between academics and practitioners are also rare.

Research and publishing are among the most misunderstood of academic responsibilities. Most practitioners have no appreciation for the time many academics take for these activities, or for the ardor with which they pursue them. Practitioners are even less aware of the pressure on academics to have a research agenda, a central hypothesis toward which all of their research needs to point. Further, practitioners do not understand pressures in academia to focus attention toward the theoretical and away from the practical. An article or monograph entitled "How to write online help" will be frowned on by their university evaluators; if the same article is entitled "Implications of differing rhetorical strategies in electronic instructional and informational texts for diverse audiences," it will be received more favorably. Academics, then, are pressured not to conduct research and write articles that will have an immediate value to practitioners; rather, the articles should help resolve an overall research hypothesis (which should be theoretical rather than practical) and should avoid concerning itself with everyday problems of the world of work (Bosley, 1995). One academic–business collaboration failed because practitioners believed that the academic's proposed publication would give away the business advantage inherent in the research results. Even though an academic insisted that the publication would deal exclusively with theoretical implications of the research and not with specifics, the study was never done.

To many practitioners, scholarship in the field is invisible (Hayhoe, 1998). They may belong to the Society for Technical Communication (STC) and read its journal, *Technical Communication*, but they may be unable to name any of the more scholarly publications in the field. Because they do not participate in scholarly discussions going on in the field, they do not see the results of academic efforts. This is one factor leading them to believe that academics do not work as hard as they do. It can also engender a level of distrust and even contempt for work that could involve collaboration with academics. Some academics, in turn, resent the fact that so few practitioners participate in the scholarly conversation. Some believe it best not to attend to the world of work at all. They consider it ideal to do their research and publishing without any concern about whether it will be of immediate (or long-term) benefit to practitioners. Secure with the thought that technical communication is now a scholarly discipline in its own right, with degree-granting programs at many institutions and its own body of literature, some technical communication faculty follow the lead of other academic disciplines and do not concern themselves with whether their research and publication has any practical utility. This view discourages conversations and collaborations with practitioners.

POWER

Even though professors are rarely fired and have long-term power associated with affecting what people learn and how they think, they have very little power as individuals and less than they think they do as groups. In contrast, practitioners at many companies can amass considerable individual and financial power. Even those at the lowest levels can be empowered to spend amounts of money and use resources that professors can only dream about. Midlevel managers in business have considerably more financial and personnel-related power than anyone in academia. Because academia continues to function as a bureaucracy, even the most piddling financial expenditures are carefully recorded and tracked. This discrepancy in the ability to commit resources quickly and decisively can derail collaborative efforts, especially those that present themselves as short-term, quick-turnaround opportunities where both groups must make rapid, firm commitments.

Similarly, decision making in academia is expected to be a collegial, collaborative enterprise. Power is feared and thus is spread among many. Much of the "overhead" time spent in meetings in academia is the result of power being distributed so that no one has too much. Because academics do not trust power, they require themselves to spend many hours a week exercising (and fighting over) their pieces of it. In contrast, although businesses are beginning to work more on collaborative models, power is still concentrated in a group of managers who can make decisions firmly and spend money quickly.

Mutual frustration is inevitable due to differences in power between academia and business. The academic mistrust of power and tendency to disperse it clash with the business tendency to make decisions and move toward fast, efficient results. Practitioners easily become frustrated when academicians want to have more meetings to discuss the subject at hand in ever more detail or when academics must seek approvals at ever higher levels of bureaucracy. Academics, distrustful of one person making decisions too quickly and insisting on considering all aspects of a question before reaching a conclusion, cannot understand practitioners' insistence on jumping into action and on their requirement for rapid results (Johnson-Eilola, 1996).

For academia, the long-term implications of prolonged decision making are frightening. Because no one can act quickly, it is impossible to adapt and change quickly. Whereas businesses are increasingly realizing the need to redefine them-selves and change quickly in response to society and technology, academia resists rapid change. Although much of the curriculum in academia is not subject to the latest trends in society and technology, the fact is that the subject matter, pedagogical methods, and, indeed, professors themselves seem woefully out of date to many students. With the Internet and other technologies moving into traditional academic territory, traditional ways of teaching could eventually face a fight for survival. With businesses using power to bring about rapid change and academia dispersing power to prevent change, professors in technical com-munication must constantly struggle between continuing with proven, conven-tional modes of teaching and discourse or offering newer modes of interaction brought about through technology. Opportunities for collaboration could be lost if businesses hope to learn more about such new modes, while academia has trouble keeping up with them.

TRUST

There is also a significant trust issue between academics and practitioners. Some of this is the "us versus them" problem that is evidenced in all social groups, but much of it is due to significant cultural differences between the two groups. Some academicians scorn the profit-driven motives of business people and are reluctant to work too closely with them. Some business people do not trust aca-demics to understand and respect their needs, be sincere in working together toward the "bottom line," stay focused, or deliver promised results by scheduled deadlines.

PHILOSOPHY

Academics are likely to examine what they do and consider how it relates to their larger communities (city, state, nation). They tend to think of themselves as working mostly for the interests of society (students and other members of their discipline). They see their work and their discipline as larger and more important than the institution where they are employed. Practitioners tend to identify less with society or a single discipline and often view themselves first as members of a certain company. Also, whereas academics are expected to have a research agenda that drives their careers, one that entails a philosophy of some kind, practitioners rarely have such an overriding philosophy, although they may be aware of a company mission statement and keep that in mind as they work for the organization. Collaborations can be strained when practitioners care about results only insofar as they benefit their own business and not the larger society. They can also suffer when academics perceive that business people are "selling out" because they are more loyal to their companies than they are to the larger society.

A similar issue concerns the philosophy of day-to-day work. Collaboration between academics and practitioners requires one group to operate on turf where it is unsure and uncomfortable. Academics working in a business environment may be uncomfortable with the pace, the lack of time and attention to study and contemplation, and the focus on short-term goals and profit. Practitioners working in academic settings may be put off by ill-defined short-term goals, the lack of clear decision making, and bureaucratic and administrative structures that often prevent routine (much less radical) processes from being implemented and completed.

REWARD STRUCTURES

Academics and practitioners work under different reward structures. Average salaries, according to the STC and *The Chronicle of Higher Education,* are similar for technical communicators and college professors. However, the top-end salaries of business positions are considerably higher than those at the top end of the academic scale (unless one gets into administration). In general, technical communicators with a few years of experience make more money than professors with a few years of experience. Most of those who teach technical communication have watched graduates get jobs paying near (or sometimes exceeding) what they are paid. However, salaries, on the whole, do not contribute as much to collaborative difficulties as do the disparate systems of promotions and raises.

Academics who start as assistant professors can be promoted twice in their careers (thrice if you count getting tenure as a promotion). They can also go into administrative positions and departmental head positions, but most academics do not view those as promotions in the way that business people look at promotion. In business, most entry-level people expect to be promoted several times within the first few years they are working. Some companies have several pay grades at each level of employment, so that something like 10 to 20 promotions and/or pay-grade increases are possible. Practitioners work toward the next level or pay grade with the expectation that if they perform well in the short term (1–3 years), they can expect (and ask) to be moved up. Academics work toward promotion also, but they assume that it will take years and considerable publication and professional activity to do so. Therefore, in business, there is a much more immediate sense of being rewarded for good work. This disparity can lead to difficulties with collaboration, as practitioners consider the short-term implications of the collaboration, hoping that it will contribute in some way toward promotion within a few months or a year at most, whereas academics tend to consider the positive implications over a longer term. For example, academics may be willing to pursue a study that would lead toward publication over several years, whereas practitioners would rarely be willing to do so.

Nonpecuniary reward systems differ also. Academics buy some prestige at the expense of income. They are *Dr.* So-and-so, a *professor*. In a society that allegedly does not have titles, most academics have at least two (more if they are chairs or deans). These bring prestige from the external society, although not so much from within the institution (prestige within the institution comes more from research and publishing than from teaching). In contrast, prestige at a business comes from management success or from fulfilling some of the company's missions better than others do, from becoming a corporate hero. Management literature in the 1980s and 1990s was full of examples of corporate heroes and how they should be encouraged, paid well, and retained. Today, businesses still have stories of corporate heroes, and managers are being trained to tell stories about them to their practitioners, as a way of spreading by example the desired employee traits in the company culture. This goal is also supported by award programs and incentives such as bonuses, team awards, and president's clubs.

Academic rewards, then, are viewed as long-term investments. Academics do not mind working on projects that have only a long-term payoff. In fact, many assume that their research and publishing projects may take them years. Practitioners, on the other hand, are constantly striving to get to the next pay-grade increase or promotion. They constantly regard every demand on their time in relation to how it will affect their next move up. They often see collaboration with academics on long-term projects as a waste of time and as requiring too much time for the amount of short-term payback that they will receive. Indeed, practitioners, particularly those in technical communication, are often so connected to the fast-paced, short-term, develop-and-deliver cycle that they cannot even contemplate the possibility of collaborating with someone in academia (Bosley, 1995), particularly if it requires a long-term commitment.

DISCUSSION

Operating in these two different worlds can be like traveling between countries. One must change languages, customs, philosophy, thought patterns, and motivations. Significant differences in the perception of information, language and discourse practices, collaboration models, employment terms and philosophies, and reward structures all contribute to making collaborative efforts between academia and business difficult. Yet, because academic programs and the practice of technical communication are mutually dependent, for both groups to succeed, they must communicate and collaborate effectively. To do so, they must acknowledge and overcome cultural rifts that stand in their way.

Other disciplines have succeeded in establishing collaborative working relationships, and we in technical communication have found some ways to do the same. The type of collaboration that seems to work best is that which transgresses the fewest of the cultural divides. That means, in general, that successful collaborations will be short-term, will provide benefit for both sides, and will not challenge either culture's basic principles. One example is the internship, which perhaps is the most common and popular collaborative effort (Bosley, 1995; Hayhoe, Stohrer, Kunz, & Southard, 1994; Southard & Reaves, 1995). Internships are generally short-term, measured in weeks or months. The business benefits from getting the services of a fledgling technical communicator for free or for a nominal salary. They further benefit from the possibility of discovering, training, and wooing a potential future employee. They might also benefit from learning something about the latest research or scholarly ideas that the student brings to the job. Academia benefits from enhancing its goal of improving the student's education and better preparing the student for work after graduation. Neither side has to compromise any of its basic tenets.

Another example is usability testing (Howard, 1999). Business can send products to technical communication classes to be tested, again for free or for a nominal cost compared to contracting with a professional testing firm. They get the benefit of having testing done that they otherwise would not be able to complete, and they can use the test results to enhance the usability of their products, thus improving sales and reducing service costs. Academia again benefits from having their students work on real-world problems rather than simulated ones, and better preparing students for their lives after graduation. These types of projects are short-term, do not cross any of the more serious cultural divides, and benefit both sides.

Even shorter term collaborations of only a few hours can succeed. One consistently successful pedagogical method is to have practicing technical communicators speak to classes about the documentation functions within their organization. Another is to take technical communication students on tours of the publications departments in nearby organizations. A third approach is to hold joint meetings of classes with STC's regular monthly meeting. Although difficult to schedule, these joint meetings provide a superb opportunity for collaborative communication, exchanges of ideas and concerns, student learning about possible jobs, and employers learning about possible future employees. All of these short-term collaborations do not confront the more serious cultural divisions.

More problematic are larger scale projects such as collaborative research and design efforts. Projects that require longer periods of time often provide no short-term benefits to either side, which is especially problematic for the business partner. They can require more project management activities, with schedules and deadlines that can cause friction. They can also require much more communication, which often exacerbates language and discourse differences. And they frequently lead to a conflict at the conclusion, when academics want to publish results and practitioners want to keep them proprietary. For collaborations of this type to work, it is mandatory that expectations and results be carefully defined at the outset, including designation of specific milestone dates and detailed descriptions of what will happen on those dates. Both sides must understand from the start exactly which outcomes will be considered proprietary and which can be published. Furthermore, commitments for resources, people, time, and finances must be carefully defined at the onset. In short, the groups must consider each of the possible cultural impediments to their desired collaboration and find ways that, if not avoiding them altogether, can at least mitigate their effects.

As we have seen, collaboration between business and academia is essential for both groups to succeed. Therefore, technical communication specialists must overcome their cultural differences so that they can find ways to make collaboration projects both possible and mutually beneficial.

REFERENCES

Barchilon, M. G., & Kelley, D. G. (1995). A flexible technical communication model for the year 2000. *Technical Communication, 42*(4), 590–598.

Bosley, D. S. (1995). Collaborative partnerships: Academia and industry working together. *Technical Communication, 42*(4), 611–619.

Hayhoe, G. F. (1998). The academe–industry partnership: What's in it for all of us? *Technical Communication, 45*(1), 19–20.

Hayhoe, G. F., Stohrer, F., Kunz, L. D., & Southard, S. G. (1994). The evolution of academic programs in technical communication. *Technical Communication, 41*(1), 14–19.

Howard, T. (1999, March). *Creating academic/industry partnerships through usability testing projects.* Paper presented at the Association of Teachers of Technical Writing Conference, Atlanta, GA.

Johnson-Eilola, J. (1996). Relocating the value of work: Technical communication in a post-industrial age. *Technical Communication Quarterly, 5*(3), 245–270.

Kunz, L. D. (1995). Learning and success in technical communication education. *Technical Communication, 42*(4), 573–575.

Southard, S. G., & Reaves, R. (1995). Tough questions and straight answers: Educating technical communicators in the next decade. *Technical Communication, 42*(4), 555–565.

Tebeaux, E. (1996). Nonacademic writing into the 21st century: Achieving and sustaining relevance in research and curricula. In A. D. Hill & C. J. Hansen (Eds.), *Nonacademic writing: Social theory and technology* (1st ed., pp. 35–55). Mahwah, NJ: Lawrence Erlbaum Associates.

2

Jumping Off the Ivory Tower: Changing the Academic Perspective

DEBORAH S. BOSLEY
University of North Carolina at Charlotte

Technical communication faculty strive to develop working partnerships between technical writing practitioners and academics and to influence practice. Much has been written about the need to strengthen ties to industry, develop joint research projects, and collaborate in developing internship and learning opportunities for students (Bosley, 1992; Hayhoe, 1998, 2000). However, in general, except for the few academics who have developed long-standing relationships with practitioners, such programmatic partnerships are still relatively rare. What stands in the way of such partnerships?

My contention is that academics themselves often set up barriers that militate against such partnerships. These barriers exist, in part, because of academics' tendency to focus on differences between academia and industry rather than on similarities of their work environments, work practices, and workplace documentation processes and products. In addition, academics tend to be influenced by traditional assumptions about self-definition and, through behavior based on those assumptions, they tend to separate themselves from practitioners in unproductive ways. Finally, academics tend to underestimate the value of their research as an aid to practitioners. As a result, they inadvertently keep such research results within the academic community instead of disseminating them to practitioners.

In this chapter, I first delineate similarities by describing common ground that both technical communication academics and practitioners share. I then suggest ways for academics to make perceptual and behavioral changes that are essential to influencing industry practice. Finally, I offer suggestions on how to capitalize on those changes by extending both the boundaries and the definitions of academic expertise—all of which, I believe, will extend the influence of technical communication academics beyond the university. Just as practitioners influence academics by giving them insight into practical experiences in industry, academics have the potential to influence the frameworks that practitioners use to think about their choices, styles, and methods, as well as their products and practices. This mutually reciprocal influence will succeed, however, only when academics and practitioners understand that there is important common ground that both groups share.

THE COMMON GROUND

Most academics (and practitioners, I might add) believe that academia and industry represent totally different philosophies and, as such, share little common ground. Academia has a long-standing tradition of viewing its work environment as substantially different from the world of work outside "ivory towers." Dicks (chapter 1, this volume), for example, points to cultural differences. From this point of view, it is inevitable that these different cultures will challenge industry and university collaborations, but these challenges are not insurmountable. Nathan S. Ancell (1987), chairman of the board at Ethan Allen, Inc., suggests that common goals of business and education do exist. He points out that

> ... both communities have distinct missions and methods. Elements of those missions and methods are compatible and, indeed, complementary. However, the ability to recognize and utilize that common ground has not always been realized on the campus or in the work place. In the past, university faculty members have at times feared contamination of liberal arts education by the business perspective and involvement [while] business people found the ivory tower perspective of academics to be somewhat fuzzy and unrealistic. (viii–ix)

Thus, academics and practitioners alike recognize that the "missions and methods" of both communities are critical to the success of both.

BREAKING NEW GROUND

Despite the plethora of articles about our differences (Bosley, 1998; Hayhoe, 1998) technical communication faculty and practitioners share many more similarities in work environments, work practices, and workplaces documentation processes and products than are currently recognized. Academics tend to overlook similarities that exist between the numerous activities that they do admin-

TABLE 2.1
Similarities of Academic and Industry Work Environments and Practices
in the Field of Technical Communication

Concept	Academic	Industry
Collaboration	Interdisciplinary, departmental, and university committees.	Cross-functional teams, collaborative documentation teams.
Text and visual products	Proposals, grants, reports, procedures, white papers, user manuals, tutorials, online material, memoranda, meeting minutes, letters, promotional and public relations material.	Proposals, grants, reports, procedures, white papers, user manuals, tutorials, online materials, memoranda, meeting minutes, letters, promotional and public relations material.
Foundations of expertise	Primarily theory with some application.	Primarily application with some theory.
Need for research	Often need and seek research results to substantiate theory and pedagogy (practice).	Often need and seek research results to substantiate practice, but little time to sift through academic publications to find relevant research information.
Administration/ management	Hierarchy, clear-lines of reporting: chairs, deans, presidents, chancellors, etc.	Hierarchy, clear-lines of reporting: supervisors, line supervisors, managers, presidents, CEO, etc.
Education/training	Constant need to keep up with changing technologies that influence technical communication practice.	Constant need to keep up with changing technologies that influence technical communication theory and practice.
Status	Feel lack of status, often marginalized within departments.	Feel lack of status, often trying to prove value to products/ services.

istratively and programmatically and practitioners' work. Table 2.1 illustrates common ground shared by academia and industry in terms of communication skills necessary to succeed in both arenas. Each category gives examples of similarities of practice.

If technical communication practitioners and academics were able to recognize many of these similarities, perhaps the gulf that so frequently separates them would shrink. By recognizing common ground among many of the conditions under which we work and write, both groups might better understand each other.

Collaboration

Universities have always conducted institutional business (to a point) via committees. Collaborating with colleagues is the primary way decisions are made that involve faculty and curricula. The advent of interdisciplinary degrees, particularly in the field of technical communication, requires that faculty across different curricula cooperatively design and administer such programs. Similarly in industry, primarily as a function of the Total Quality Management (TQM) movement, and more recently quality assurance or process management initiatives, cross-functional and collaborative teams are predominant modes of work production. In industry and in academia, teams and committees are necessary for administering or managing departments and units, so it is clearly necessary for both academics and practitioners to learn to cooperate extensively with their colleagues. Academics and practitioners, therefore, share both the experience and the frustrations of teamwork.

Text and Visual Products

Both universities and corporations focus much of their energy on "the bottom line." Industry seeks profit, but to a large extent (particularly at major research universities), the ability of the university and its faculty to sustain themselves depends on the quality and quantity of research. In both academia and industry, technical communication specialists contribute significantly to the accumulation of their own bottom lines through the production of well-written documents.

Within a university environment, faculty are constantly preparing documents that are the lifeblood of their organization. Although such documents generally have no relationship with products or services as they do in industry, proposals, grants, and reports are often tied to a faculty member's ability to secure funding for research, teaching projects, or both. If the "currency" of the university is shared knowledge, then such documents clearly help generate the means to produce and disseminate knowledge.

Foundation of Expertise

Academia clearly valorizes theory above practice. In fact, many faculty are openly discouraged from focusing their research on practice. However, by its very nature, a technical communication curriculum must blend both theory and practice. Preparing students to become technical communicators means teaching them the "why" of document design as well as the "how." Teaching theory without practice leaves students lacking in some of the concrete skills they need to be successful practitioners; teaching practice without theory produces students who lack a repertoire of conceptual models and frameworks to draw on when they encounter the complex requirements and relationships inherent in most workplace communication situations. Therefore, a careful blend of theory and practice is essential to prepare students and practitioners for the world of work. Despite the general sense among practitioners (see elsewhere in this chapter for survey results) that theory and research are useless, in actuality, theory often enables practitioners to argue for (or against) current documentation practices and to propose new ones. As a counterpoint, a heavy dose of practical experience is critical in helping students understand the "how" that supports theoretical expertise in technical communication and negotiate the constraints of work budgets, schedules, resources, and environments.

Administration/Management

Although cooperative or collaborative teams and committees exist and serve important functions in both academia and industry, hierarchical structures still dominate in both environments. Who makes critical decisions remains a function of the management philosophy and structure in both arenas. Despite faculty senates or councils and the perceived reliance on faculty decision making, most decisions essential to the future of the university are made at levels far beyond the faculty. Practitioners also often work in environments that still maintain hierarchical structures, and decisions generally are made in a top-down fashion. Both groups are often frustrated by their lack of control and decision making.

Education/Training

The proliferation of business books about "the learning organization" and the billions of dollars spent annually on employee training attest to the emphasis on life-long learning within industry. Even the Society for Technical Communication (STC), which currently has a membership of over 24,000 practitioners and academics (though the latter are, in fact, underrepresented) aims to help technical communicators continuously add to their knowledge of the field. The increasing number of technical communicators that attend conferences, receive training, and seek higher degrees indicates this emphasis on the importance of continued learning and professional development. Many organizations routinely pay for continued education. TIAA-CREF, the largest teachers' pension fund and a competitor in the investment industry, for example, pays college tuition for any employee who wants to take college courses or seek a degree.

Obviously, academics are also devoted to life-long learning. Technical communication faculty, for example, often take workshops to keep up with advances in computer technology.

Status

Technical communication academics often feel a lack of status. One problem is a lack of understanding about the field among university administrators and colleagues. For example, Pratt (1995) (former president of the American Association of University Professors) suggests that ". . . the enterprise of 'business and technical writing,' . . . largely enables students to write better memos, presentations, and letters of introduction" (p. 36). This attitude toward technical writing is, unfortunately, still pervasive in academia. Most technical communication programs have had to fight for recognition within their university, often within their own departments. Many programs exist within humanities (often English) departments that are still dominated by faculty in traditional areas such as literature, who tend to disdain technical communication programs with their focus on professionalization. Tension in academic departments can escalate when colleagues perceive only the instrumental nature of the field and lack understanding about its rhetorical foundations.

At the same time, practitioners also feel a lack of status within their organizations. For example, many documentation experts resent that despite serving as user advocates, they are brought onto R&D teams well after a product has been developed. Technical communication practitioners also tend to face corporate layoffs sooner than others, such as product designers and engineers, whose contributions are perceived as more central to company goals. Suffice it to say that this lack of status frustrates and weighs heavily on both groups.

PERSPECTIVES

In addition to recognizing the common ground that they share with practitioners, academics also must understand traditional assumptions or perspectives that drive much of the academic side of the field of technical communication. In their desire for acceptance and influence, academics can adopt strategies to change both their image and their perceived usefulness to business, industry, and universities. Two of these traditional perspectives are described next, followed by new perspectives, both to help practitioners understand academia in new ways and to help academics recognize what influences they may have on practice.

Traditional Perspective: Technical Communication Faculty Primarily Define Themselves as Teachers and Researchers and Seldom as Practitioners

Many academics in technical communication define themselves primarily as teachers or researchers. Secondarily, they consider themselves professionals and members of a larger technical communication community beyond the university. Some do consulting work. Many belong to STC, the largest professional organization of practitioners, and some also join other professional organizations such as The Institute of Electrical and Electronics Engineers (IEEE), Special Interest Group for Documentation (SIGDOC), Women in Communication (WIC), and the American Society for Training and Development (ASTD), which all include more practitioner than academic members. However, even within the context of these and other organizations, professors tend to present themselves as academics.

By defining themselves primarily as teachers, academics tend to feel strongly that one of their most important responsibilities is to educate future technical communication professors and practitioners. Doctoral program graduate students sometimes take industry jobs but mostly aim for academic careers, whereas bachelor's and master's degree program graduate students mostly become practitioners. All these programs are judged, at least to some extent, by the quality of their graduates.

Many technical communication professors also define themselves as researchers. Particularly for those who work at 4-year universities, the quality of their research can determine whether they receive tenure or promotion. The pressure for them to publish ("publish or perish" still applies in most universities) often takes precedence over most other activities.

In addition, technical communication faculty often define themselves as administrators. Many run technical communication programs. Note the healthy existence of the Council of Programs on Technical and Scientific Communication (CPTSC), an organization focused on concerns of program administration. As administrators, they typically manage other faculty, produce a variety of documents associated with administrative work, and carry out business within the departmental and university hierarchy. Even within this role, however, faculty tend to view themselves as academics, not as practitioners.

New Perspective: Technical Communications Faculty Should Define Themselves as Practitioners Both Within the University and Within Local Communities

Unfortunately, although academics rarely define themselves as practitioners, they often attempt to influence the practice of technical communication. Doing so can cause some resentment, especially when academics have never held jobs in nonacademic contexts and lack a full understanding of workplace practice. If faculty hope to influence practitioners, they must first begin to define themselves as practitioners and engage in technical writing practice whether within the university or in more traditional workplace environments. Without such credibility, academics should not be surprised when practitioners ignore what they have to offer.

Most technical communication academics and practitioners could fill volumes with stories of their lack of status and the underappreciation of their value. Changing university and corporate culture is, of course, extremely difficult, but changing the status of technical writers within academia and industry may be an easier sell. Yet, to improve the status of technical communicators, it is critical that academics begin to define themselves as practitioners. I contend that once practitioners find out that professors have had actual experience as technical writers, they are more likely to respect them and what they have to offer. To develop this new self-definition, faculty might consider seeking ways to write documents that are crucial to the well-being of their university.

Let me illustrate this point with a few examples from my environment, the University of North Carolina at Charlotte. Several years ago, the former chair of the English department was asked to recommend someone who could help the administration write the report for the upcoming National Collegiate Athletic Association (NCAA) accreditation visit. They asked for someone who was a good writer, and my chair recommended me. Because I had worked on a variety of university committees for years, faculty and administrators across campus knew and respected me as a teacher and scholar. However, they were looking for someone who could also function as a practitioner. Accepting this definition (and being paid for it!) put me in constant contact with the chancellor, the provost, the NCAA representatives, highly valued faculty from across the university, and business people who also served on the committee. When the NCAA committee wrote its findings, mention was made of how well we wrote our report. During most of the meetings, administrators would refer to me as Dr. Bosley, or Deborah, "our resident technical writer." Because of that experience, I often am asked (from the provost's office down through the administrative hierarchy) to help write or edit important university documents, critique university Web sites, or train faculty in communication issues related to distance education.

Defining myself also as a practitioner has increased my status within the university. At no time have I been asked to write documents or train faculty without compensation, perhaps because they define me (in some situations) less as a professor and more as a practitioner. But the real benefit to me as a professor has been working on cross-functional teams, meeting deadlines (even when I knew time was overriding quality), learning the language of whatever writing context I was engaged in, and understanding the politics of the worksite and how those politics impinge on documentation production. These experiences have given me insight into some demands on practitioners that I would not have understood were it not for these situations.

Faculty also can develop this new definition by seeking ways to write documents that are crucial to the well-being of their local community. In addition to defining themselves as practitioners within the university, technical communication professors should find ways to become practitioners in the business community. Doing so is a way to make obvious their expertise and value. Again, by working as practitioners, faculty can increase their credibility among practitioners.

For example, the business community in Charlotte, North Carolina, is extremely involved with the local public school system. As a result, most public school committees are filled with business volunteers. In an effort to improve communication between the schools and the public, I volunteered to rewrite the procedures for applying to magnet schools. By doing so, I was invited to attend several meetings to present my objections to the original document as well as get input from a variety of "publics" on the purpose, the audience, and the content for the rewritten form.

Although I was introduced as "Dr. Deborah S. Bosley," it became clear that I was participating as a technical writer, not as a professor. During a local STC meeting, when I talked with practitioners about this experience, one stated, "It's important to have technical communication faculty functioning as technical writers. Now you can better understand what our working lives are like." Finally, I have contracted with several corporations and agencies to produce documentation, revise forms, and critique Web sites. In each situation, my experience as a "contract writer" has increased my understanding of practitioners' tasks.

In addition, faculty can seek ways to engage students as practitioners, thereby increasing student engagement with business people and members of the university and external community. Many academics have long argued for the value of project-based learning in which students are engaged in writing or communications that involve purposes, audiences, and contexts beyond the university. The exigencies created by such writing assignments teach students about corporate cultures in ways that text-based case studies cannot. Developing projects for students can put them in contact with practitioners, business people, and others from across the curriculum. For example, at UNC Charlotte, students have written tutorials for the public library, produced promotional materials for a local neighborhood trying to improve the quality of life in their community (funded through a grant from HUD), produced patient information for a medical clinic, and revised material produced by UNC Charlotte's computing services unit. In each of these cases, students interviewed the audience, learned about the politics of the environment, developed project management skills, learned about the continual struggle between quality and timeliness, participated in business or agency meetings, and then brought their experiences back to the classroom. When these students seek employment as technical communicators, these real-world experiences as practitioners give them a kind of credibility unavailable to students who produce technical documents solely for the classroom environment.

Overall, faculty need to take the initiative to learn much more about the work environment of practitioners. Only by understanding the constraints under which practitioners work will academicians be able to provide assistance and expertise of value to practitioners.

Traditional Perspective: Academic Research Need Not Have Any Practical Application and Can Be Reported Primarily in Academic Journals

Practitioners and academics also hold differing views on research and the dissemination of research results. According to Carliner (2000), academic research can be categorized in two ways. Evaluation research is intended to provide people with information on which they can base practical business decisions ... business people specifically expect that research will provide definitive answers to questions for which there are no definitive answers, such as the ideal organization of a technical communication department or the ten things that guarantee quality. (p. 550)

Original research intends to "further the body of theory [and knowledge]. [Academics] often expect people in industry to value this research for its own sake, regardless of whether people in industry can see the relationship of this research to their own practice" (p. 551). As Gery suggests:

To professors, practicing professionals often seem to take a perfunctory approach to learning. Practicing professionals seem more interested in desktop publishing and online authoring than in communication skills. When they do learn communication techniques, practicing professionals frequently do not want to understand why to use certain techniques, just how to apply those techniques in 10 steps or less. . . . If only they were willing to spend the time learning the "why" underlying the "how," they could have avoided the resulting disaster. (p. 5)

Many academics have to focus on this second kind of research if they are to succeed in a university environment (tenure and promotion). In fact, research that adds "theory" to the body of knowledge about technical communication is still valorized in most university environments above research with more practical applications. Academics in universities conduct research from both an intrinsic curiosity and the extrinsic reward system that values research above all other academic activities. Whether the research has practical application is generally irrelevant.

This problem can be illustrated by the situation of a colleague in a department of sociology. For years, he has conducted and published significant research on the relationship between alcoholism and domestic abuse. Although his research could directly or indirectly save lives, his department culture made it clear that publishing such research in trade publications would diminish his value as an academic researcher. In another situation, a technical writing faculty member wanted to take a leave of absence to enter a corporate environment to conduct research on corporate document design. His dean indicated that it was hard for him to justify to university committees why someone (in this case) from an English department needs to spend any time in industry.

It is no wonder, then, that such research often has little if any influence on practitioners. Unlike practitioner researchers, academics rarely do research to develop products that increase profit. However, academics may underestimate the value that practitioners place on high quality research—both for its extrinsic value of improving products and processes, and for its intrinsic value of introducing new ideas and concepts. Practitioners often are just as excited and motivated by new ideas as are academics.

New Perspective: Academics Should Conduct Research and Disseminate Research Results in a Manner That Practitioners Find Useful

Let me illustrate a common clash between the university classroom and "the real world." A colleague in a department of marketing relayed a story of her upper division marketing class in which students were assigned to read a research article on marketing strategies as applied to selling automobiles. One class member, an older student who had many years of experience selling cars, told her that not only were the research and subsequent conclusions irrelevant in terms of helping him and others sell more cars, but also, the data were simply wrong. The author of the article had collected statistical data, but had never talked directly to anyone who actually sold cars. The practices outlined in the article did not tally with this student's 11 years of experience selling cars. On this basis, he (and the other students in the class) found it easy to discount the results of the research and the points made in the article.

One of the most significant opportunities that academics have to influence and assist practitioners is through research. However, many practitioners find the current spate of academic research to be just as irrelevant to their needs as the earlier story illustrates. Regardless of whether they have degrees in technical communication or are practitioners based on trial-and-error and experience, practitioners are rarely influenced by research articles written by academics. Why? What is it about our research that makes it unusable by many practitioners even though many research articles appear in STC's *Technical Communication,* the journal most likely to be read by practitioners? The answers may be found in the following assumptions:

Academic Articles Are So Theory-Based That the Practical Application Gets Lost. With few exceptions, research results are often focused far more on the theoretical implications than on practical applications. Despite the convention that technical communication research articles include implications for teaching, practice, or both, practical applications are often an afterthought rather than the focus of the article or the research itself.

Academic Research Is Not Accessible to Practitioners. Practitioners rarely have access to academic journals (except, perhaps, STC's *Technical Communication*). Many read trade publications relevant to their area of business or industry. Document design articles in such trade publications are rare. If research results are to be made available to practitioners, then publishing in trade journals must become a goal for academics.

Academic Articles Are Written in a Way That Makes Comprehension Difficult. Look, for example, at the following brief description of a technical writing training program. Notice how difficult this description is to understand (the passive voice dominates).

> **Class Description**: A course in written technical communication skills focusing on the implementation of the structure and expression of language and syntax in a multitude of technical writing genres. Given this particular focus, the participant's ability to develop, apply, and illustrate an understanding of various theories and their functions and rhetorical values and responsibilities, both professionally and to an audience community, is highly stressed. Therefore, document design principles, the value of the revision cycle, and critical cognitive capabilities are main course considerations. Participants will be introduced to rhetorical theory as it applies to the workplace.

My own writing (Bosley, 1994) has suffered from the same convoluted style. Here is an excerpt from an article on the topic of gender and technical communication:

> In the same way that feminists have looked at the epistemology underlying science, technical communication as a discipline with its roots in scientific, logical positivism, has come under scrutiny for its masculinist predispositions and biases by "defining itself as the objective transfer of data, truth, and reality" [1, p. 358]. Technical communication, like scientific discourse, carries with it an implied, and often stipulated, epistemological and stylistic objectivity.

This latter example of academic language establishes the authority of the author, lends credibility to the research and conclusions, and appropriately meets expectations of an academic audience. However, this kind of language is inappropriate for a practitioner audience because it relies too heavily on presumptions and assumptions of an academic audience. For a practitioner who has neither the time nor the experience to commit to the amount of thinking and background required to read this article, the research and the conclusions—however ultimately relevant to their needs—appear irrelevant and too loaded with theory to be practical. Even one of the biggest training institutes advertises that its workshops do not "contain theories or concepts, but rather *usable* information." [emphasis added] This quote sums up the attitude that many practitioners have about academic research.

To influence and provide useful assistance to practitioners, academics must conduct research that is helpful to them, and publish the results in forms and publications that practitioners read. Clearly, the kinds of research that are valued in academia are different—not necessarily in kind, but in form—from those valued in industry. For example, most universities primarily value research and scholarship that advance theory. Research focused on improving practice or application often is, in general, anathema to promotion and tenure committees. Therefore, much research that academics conduct is highly theoretical or produces results that may be of little or no interest to practitioners.

Beyond that problem is one of dissemination. Some academicians have conducted research that speaks directly to practitioner needs. However, they tend to publish these results in journals rarely read by practitioners. Instead, academics could craft research results into short articles for business magazines, business sections in local newspapers, and cutting edge technology journals like *Wired*. For tenured academics, such publications (although certainly not valued by academics in general) remind practitioners that academic research, knowledge, and experience can be of value to them. Academic writers are capable of covering both the "how" and the "why" in clear, direct prose. For example, a technical communication colleague at UNC Charlotte once worked at a Fortune 500 company. At that time, his plant manager (who knew he was a writing professor) assigned him the task of rewriting theoretical research in a more palpable form to assist employees in developing communication skills. These kinds of publications and writings by faculty can help practitioners access, understand, value, and use our research.

CONCLUSIONS

This chapter has identified common ground shared by technical communication academics and practitioners. It also illustrates ways that technical communication academics can assist and influence the work of technical communications practitioners. I leave it to my practitioner colleagues to discuss ways in which they can influence academia. It is only through this kind of recognition—that each community has something to offer the other—that technical communicators will truly respect each other and want to collaborate and partner together for life-long learning.

REFERENCES

Ancell, N. S. (1987). *Exploring common ground: A report on business/academic partnerships* (pp. viii–ix). Washington, DC: American Association of State Colleges and Universities.

Bosley, D. S. (1992). Broadening the base of a technical communication program: An industrial/academic alliance. *Technical Communication Quarterly, 1*(1), 41–58.

Bosley, D. S. (1994). Toward a feminist theory of visual communication. *IEEE Transactions on Professional Communication,* December, 122–133.

Bosley, D. S. (1998). Collaborative partnerships: Academia and industry working together. *Technical Communication, 42*(4), 611–620.

Carliner, S. (2000). Finding a common ground: What STC is and should be doing to advance education in information design and development. *Technical Communication, 42*(4), 546–554.

Gery, G. (1995). Preface to special issue on electronic performance support systems. *Performance Improvement Quarterly 8*(1), 3–6.

Hayhoe, G. F. (1998). The academe–industry partnership: What's in it for all of us? *Technical Communication, 45*(1), 19–20.

Hayhoe, G. F. (2001, March). *Inside out/outside in: Transcending the boundaries that divide the academy and industry*. Keynote address at the annual conference of the Association of Teachers of Technical Writing, Denver, CO.

Pratt, L. R. (1995). Going public: Political discourse and the faculty voice. In M. Berube & C. Nelson (Eds.), *Higher education under fire: Policies, economics, and the crisis of the humanities* (pp. 35–51). New York: Routledge Chapman and Hall.

3

Researching a Common Ground: Exploring the Space Where Academic and Workplace Cultures Meet

ANN M. BLAKESLEE
Eastern Michigan University

Some recent research by professional writing scholars points to differences between academic and workplace cultures and their respective genres. Freedman, Adam, & Smart (1994), for example, show how writing performed in the classroom, specifically case study writing, differs in significant ways from workplace writing. However, whereas these scholars point to differences in the two cultures, Russell (1997) explores and theorizes potential connections between them. He says the genre system of a classroom "forms a complex, stabilized-for-now site of *boundary work* [his emphasis] . . . between the discipline/profession . . . and the educational institution" (p. 530). Russell also says that intertextual links with genres in the profession make the classroom genres that students write resemble those of professionals.

We have not, in our empirical research on genres or workplace writing, explored the overlapping space to which Russell alludes. In this chapter, I argue for shifting the focus of this research to call greater attention to this space. I suggest that doing so could serve as one means of bringing the concerns of academia and the workplace closer together. If we acknowledge that these worlds are distinct cultures, which both academics and practitioners tend to do, then it is important to study how these two cultures and their respective genres differ.

Some chapters in this collection address precisely these differences (see Dicks, chapter 1, this volume). However, it is valuable, as well, to acquire a better understanding of the overlap between the two cultures because our similarities, more than our differences, may enable us to identify and address our shared concerns. My argument in this chapter thus rests not solely on assumptions of difference but also on metaphors of common ground, overlap, and similarity. I acknowledge the differences between the cultures, but I contend that they share an important feature, the goal of affecting a world that is rapidly changing as it is shaped by new information technologies. For this shared goal, both cultures seek to influence the consequences of these technologies at the point of entry at which they are conceived and designed. Finding and focusing on a common ground—or overlapping spaces—may allow the two domains to do a better job working together as we seek to stake a claim to this work. One point on which members of both the academy and industry agree is that this claim falls rightfully and predominantly to technical communication more than to other fields.

I argue in this chapter that we should work to develop a program of research that helps us to locate, define, and enhance this shared ground. Furthermore, because some of the main dimensions of culture are communication structures, idioms, and genres, investigating the respective communication structures and genres of the two cultures may be a good way to begin searching for an overlapping space. Unfortunately, nothing in the field of technical communication deals specifically with the kind of research that may help academic and industry professionals grasp the structures and idioms of communication that constitute a shared meeting ground.[1] Therefore, I propose ways to research where and how these structures overlap. I also show what this research might look like by abstracting from immediate case studies of classroom–industry collaborations; demonstrate how the collaborations, along with research that both addresses and results from them, may begin the foundation for a common ground; and make a case, as well, for practitioner research as an additional means of constructing and helping us meet on a common ground.

---

[1] I should note that an alternative approach might be to arrive at a common understanding of each world's roles and efforts without feeling compelled to find overlaps. With this approach, the cultures would remain distinct, but possibly agree both on strategies and on each world's own best, but separate, contributions. I contend, however, that we stand to gain far more by making contributions to this realm jointly rather than separately. For example, by acting together we stand a greater chance of increasing our visibility and perceived value in areas that other fields may also be trying to claim. We also can be more productive as we work toward mutual goals and visions for new technologies if we share and exchange our ideas and research in more collaborative ways.

THE FOCUS OF EXISTING RESEARCH ON WORKPLACE WRITING

One purpose of academic research on professional writing and genres typically has been to improve, for pedagogical purposes, the understanding academics have of workplace writing. Academic researchers often look at and then describe the workplace, specifically its practices and genres. Some aim to convey and emulate these practices and genres in the classroom. Many of these studies end up emphasizing differences between writing in academic and workplace settings, implying, often, that the writing students do in technical and professional writing classes is dissimilar from and, therefore, of less quality than the writing they will end up doing in the workplace.

Freedman et al. (1994), for example, contend that case study writing performed in professional writing classes differs in significant ways from workplace writing and, therefore, does not fully convey important features of professional genres, such as the rhetorical contexts and the social actions and motives entailed by them. These scholars claim that case study writing reveals similarities and differences between university and workplace genres and that the social action students undertake in classroom writing is radically distinct from the social dynamics of workplace writing. In short, by looking at case study writing from the perspective of genre studies, Freedman et al. (1994) note the extent to which school writing and culture differ from workplace writing and culture. They also draw support from Knoblauch (1989), who says, "Workplace practices are embedded in additional layers of social reality and cannot be understood— or learned—apart from them" (p. 257). Knoblauch's statement underscores that workplace writing encompasses all aspects of a project in which actors (writers in these cases) engage fully in the everyday life of their organizations. Technical communication thus encompasses the social dimensions of organizational life as much as it encompasses a particular communication product.[2]

---

[2]Many scholars now are making strong arguments for getting even more at these social and organizational dimensions. Most existing research is descriptive. However, many scholars are arguing now that we need to examine, to a greater extent than we have, the effects various genres have on organizational and institutional cultures. These scholars claim that we need to move away from just describing genres textually and move more toward attending to their social, political, and cultural implications. Along these lines, Freedman and Medway (1994) argue that we "need to see genres . . . as shifting, revisable, local, dynamic and subject to critical action" (p. 15). These scholars argue for rethinking the social, cultural, and political purposes of genres and for examining the assumptions, goals, and purposes of genres that remain tacit. Luke (1994), in his series editor preface to Freedman and Medway's collection, also argues that we need a system for analyzing genres that foregrounds whose interests they serve, how they construct and position their writers and readers, and who has access to them (p. ix). Berkenkotter and Huckin (1995) argue in their work for focusing our research on both micro (interpersonal and psychological) and macro (sociological and cultural) levels of activity.

Other scholars also address and point out differences between academic and workplace writing. Many of these scholars, like Knoblauch (1989), emphasize the social dimensions of such writing. Anson and Forsberg (1990), for example, emphasize that writing does not exist independently of the context or community in which it is immersed. These scholars argue that writing is cultural adaptation and that it requires "highly situational knowledge that can be gained only from participating *in* the context [their emphasis]" (pp. 222, 228). Burnett (1996) identifies and examines several factors that she contends distinguish collaborative teams in workplace and classroom contexts. She argues that these factors contribute to our understanding of workplace–classroom distinctions. Spinuzzi (1996) emphasizes that classrooms and workplaces are distinct activity systems with distinct objectives, a finding that also is significant in my work, which I show later. Wickliff (1997), concerned with similar questions, examines the value of client-based group projects, which he concludes offer a compromise between cases and internships. Wickliff's study suggests that such projects still do not simulate the workplace; however, they do provide skills that carry over to workplace settings. Finally, Winsor (1996a, 1996b) shows how students perceive the two settings differently, including what they learn and how they act in each.

Much of this research ends up differentiating, in fairly explicit ways, workplace and academic settings. However, we also should identify and examine overlapping aspects of these cultures and of their genres, and consider these aspects as potential bridging features, markers that could bring together professionals of both worlds and provide productive strategies for communicating across them. In terms of genres, we still do not know enough about where the two worlds intersect and how they may be linked. Discovering the common ground may help us collaborate as we work toward mutual goals and visions for more user-centered technologies. Regardless of the worlds in which technical communicators reside, we all bring a broad-based experience in communications and information design that we can apply, not only during a single phase, but throughout all phases of technological development. Our contributions throughout the development process are essential to making these technologies effective. Arriving at a common language and rhetorical framework could help academics and practitioners work together, not only to determine the contributions we can make in the design and development of technologies, but also to articulate and distinguish our contributions from those that our colleagues in other disciplines might make. To realize the ways in which language and rhetorical frameworks overlap in the two worlds, I focus here on genres, language, and communication conventions.

A PLACE TO BEGIN: STUDYING CLASSROOM–
WORKPLACE COLLABORATIONS AS A MEANS
TO ARRIVE AT A COMMON GROUND

What would research look like that seeks to accomplish such goals? To begin, we should consider researching sites of boundary work between academics and practitioners, for example, classroom–workplace collaborations where students complete workplace projects provided by clients as part of an academic class requirement. I have incorporated such collaborations into my technical and professional writing classes for approximately 12 years, and I periodically have studied these collaborations to determine their effectiveness as a pedagogical approach (Blakeslee, 1999a, 1999b, 2001). Findings from my studies also have led me to speculate on ways in which such collaborations may help to bridge the gap successfully between academic and workplace contexts. In particular, my research has revealed not only differences, but also similarities and areas of overlap between our cultures and their genres.

An Example: Two Teacher Research Cases

To offer an example, I draw here on two teacher research case studies that I carried out to examine these collaborations. I conducted these studies at a large state-supported research university in the Midwest (hereafter referred to as Midwest Research University), and the other at a large state-supported teaching university, also in the Midwest (hereafter referred to as Midwest Teaching University). At Midwest Research University, I studied an upper level undergraduate technical communication class for engineering students. In this class, the students researched and recommended icons for the hard-copy and online documentation of a large engineering and technology company. This project lasted 7 weeks. At Midwest Teaching University, I studied an upper level class for technical communication majors on writing computer documentation. The students in this class developed documentation for a Web-based electronic mailing list for linguists. This project lasted 10 weeks.

When I carried out my research with these classes, my primary concern was with assessing the pedagogical value of client projects (Blakeslee, 2001). I wondered if the workplace projects carried out as part of the classroom–workplace collaborations would provide students with better, or at least different, exposure to workplace writing and activity than writing assignments derived from text-based case studies. I also wondered if the workplace projects are better tools for teaching workplace writing (Blakeslee, 2001). To investigate these issues, I carried out teacher research case studies in both classes. I interviewed all of the students about their perceptions of the realness and value of the projects; their sense of the audiences they were addressing; their perceptions of feedback they received, both from me and from the client; their level of motivation in completing the projects; their sense of the degree to which the clients valued the projects; and the difficulties they felt they encountered while completing them. I also asked students to complete questionnaires about the projects, and I interviewed the clients who sponsored the projects. Finally, I collected and analyzed the documents that both the students and the clients produced during the projects. These included the final project documents as well as interim documents such as e-mail messages and memos that passed between the clients and students throughout the projects (these included responses by clients to documents produced by the students and progress reports and updates from the students to the clients).

Although well aware of the limitations of teacher research and of the criticisms that often are leveled against it (e.g., Applebee, 1987; Hillocks, 1986; North, 1987), I chose this approach for the advantages it offered. Teacher researchers, for example, have demonstrated time and again how this research may offer contextual descriptions, encourage critical reflection and dialogue, and bring about needed change (Cochran-Smith & Lytle, 1999; Fleischer, 1994; Ray, 1992, 1996). Fleischer (1994) claims that teacher research provides insights into classrooms and students that outside researchers would have difficulty discerning. I certainly saw this potential with my own studies because of my familiarity with the projects and my access to the students and clients; however, I also was aware of the constraints this familiarity posed. As Moss (1992) notes, teachers who study their own classrooms are constrained by the expectations they bring, their implicit knowledge and assumptions, and their tendency to overlook patterns that are not unique. Ray (1996) also addresses how teachers may lack sufficient perspective when they carry out research in their own classrooms and how they also may experience conflicts between their roles as teachers and researchers. Despite these constraints, I believe that this was a productive and useful way to approach these studies and that my approach yielded useful and reliable findings.

ACADEMIC AND WORKPLACE CONTEXTS OVERLAPPING IN CLASSROOM–WORKPLACE COLLABORATIONS

From my case studies, I found that such collaborations may provide valuable insights into both the differences and similarities between academic and workplace genres and practices. One particular finding that supports this observation relates to students' perceptions of the projects as transitional experiences. For example, one student from Midwest Teaching University, in describing her perceptions of the projects, addressed their transitional qualities:

> I think the transition from the cocoon of college to the real-life world would be much harder without these experiences. . . . I see it as a perfect stepping stone, a

kind of halfway point. We're not completely on our own; we still have you, and they're not getting completely professional work. . . . (Pat, technical communication major, Midwest Teaching University, personal communication)

Another student addressed how his perception of these transitional and overlapping qualities developed as the project progressed:

As we started we definitely saw it as a school assignment. Then we thought of it not as a professional or business type thing because it's not, but something that stood on its own. . . . It's not that this is part of our job, but neither was it just an assignment for class. (Tim, engineering major, Midwest Research University, personal communication)

From their experiences, students concluded that the projects resembled and contained qualities of, but did not replicate exactly, the kinds of activities and writing they believed they would be asked to perform in a professional setting. At the same time, they also perceived them to be more than just classroom projects. As one student said,

I think that your goal was to give us a taste of what it was like—not exactly, but close to what it would be like to go through a real project like this with an employer. . . . They seemed—I don't think that it was as stressful as it would be if we had been doing it for a real client [in the workplace]. (Terri, engineering major, Midwest Research University, personal communication)

Students generally viewed these activities as transitions or stepping stones to the activities and writing that they believed they would end up doing in the workplace.

If one extrapolates a bit, the transitional qualities that the students attribute to these projects suggest that academic and workplace contexts may overlap in potentially significant ways in these collaborations. The collaborations seem to create a bridge and to exist in a kind of boundary space between the classroom and workplace—the activity systems of the classroom and workplace end up overlapping with the projects (Russell, 1997).[3] Thus, the students' comments and perceptions suggest that these activities are located, not in either domain exclusively, but in a transitional or boundary space between the two domains. This boundary space suggests an area in which language, rhetorical aims, and work processes might be held in common, or might at least come into dialogue.

---

[3] According to Russell (1997), classrooms exist on the boundaries of activity systems (the university is a way into the genre system of some discipline or profession), and intertextual links between professional and classroom genres lead to similarities in the two genres (pp. 529, 531, 541). Russell also says that the closer classroom assignments get to the boundary of the activity system, the more context students discern and the more clearly they see the relationship between their own writing and that of the discipline (p. 539). Russell's work thus supports the idea of there being a shared or overlapping space between the culture of the classroom and that of the workplace, and of the value of looking at this space even more closely in our research.

It also suggests a space, either real or metaphorical, where the two cultures could meet and make progress toward achieving shared goals. Such a space may provide a productive location for examining both differences and similarities in the genres and practices of the two cultures.

An incident that occurred during one of my research projects provides an example of how this overlapping space, whether real or metaphorical, can provide a productive site for researching the genre and cultural similarities and differences of the two domains. This incident reveals the layers of social reality in which workplace practices are embedded as well as the micro and macro levels of activity in organizational life that genre theorists and researchers have addressed (Berkenkotter & Huckin, 1995; Freedman & Medway, 1994; Knoblauch, 1989).

Genre and Culture in the Icon Project

During the icon development project, students at Midwest Research University were exposed to the company's practices, mission, and products. They met several times with the clients, three writers from the company's technical communication department. They also were given a tour of the company and were introduced to several of the company's programmers and engineers. These events exposed the students to some of the more surface features of the organizational context that were immediately relevant to their work, such as job roles; workflow processes, including procedures for handing off projects; and communication channels. Exposure to these features of the organizational context seemed to impact the students as they carried out their work. They began the project as they might have proceeded in a workplace. For example, they carried out research, justified their decisions to their clients, and were held accountable for their decisions. Yet, the students were not exposed to features that are more embedded in the organizational context, such as status and authority, criteria and priorities for decision making, the flow and direction of communication, and standards for work processes and workflows. This gap in their exposure became significant and even impeded them in some respects later in the project.

About three quarters of the way through the icon project, something happened that neither the students nor the clients expected. During one of their visits to the class, the three technical writers from the company mentioned to the students, in a sort of "by the way" manner, that staff members from another office of the company, located in another state, had developed their own set of icons. The clients assured the students that these icons had not yet been accepted by the company and that the icons the students were developing would still be considered. Yet, despite these reassurances, the students began to feel that their work on the project had been for naught. For the first time, they began to view the project more as a classroom exercise than as an actual workplace project.

At first glance, this conclusion seems to have merit; however, looked at more closely, this situation reveals some of the genre and cultural similarities and differences of the two domains, especially as they pertain to language use, rhetorical aims, and work processes. First, this situation reveals a great deal about the social dimensions of organizational life, especially dimensions embedded in an organizational context that are taken for granted and easily assumed by members of the organization. These dimensions include, for example, the competition that occurs in organizations, for instance, between the technical writers who acted as our clients and their colleagues in the other office who developed their own set of icons. This situation also suggested a great deal about how hierarchy and status, another dimension of organizational context, may impact everyday activity in organizations, including their effects on the course of that activity, which may occur without much warning.

A close consideration of this situation also reveals how some of these features of organizational life existed in the classroom as well. The collaborative student teams competed with one another much like the various organizational teams. Each team developed its own sets of icons with the understanding that the clients would select the best set. Similarities also existed in regard to hierarchy and status. In the class, the students did what they were told to do by the clients, who functioned in a supervisory capacity in relation to the students. At the company, the clients had their own supervisors, and they functioned in much the same manner as the students did, as a project team doing as they were told.

Despite these similarities, numerous factors made this situation more complex—and more instructive—than it initially appeared. For example, a close consideration of this situation reveals how the politics of the organization built subtexts into the tasks the students were to complete and the requests made of them. The clients' requests to the students, and how they presented them, were influenced greatly by production and social processes of their workplace, along with their workplaces' priorities and rationales. For example, the memo the clients wrote requesting the icons spelled out very explicitly for the students the manner in which they were to develop the icons. Because of requirements that they faced themselves, the clients instructed the students to carry out thorough research, present that research for approval, develop preliminary icons based on the research, test the icons, and then arrive at and make final recommendations. The clients' concerns and requirements for the students reflected their own concerns and workplace requirements. They made it clear that only those recommendations solidly backed by research would be considered for implementation. Further, the stronger and more compelling the support, the better their chances.

The students interpreted and viewed these requirements primarily in light of their own situation; they assumed that the clients were concerned with getting a good outcome from them. What the students failed to see was how these also were organizational requirements. Because they had been exposed to and were responding to the surface aspects of the organizational context, and not to its other, more embedded aspects, they interpreted a number of the events, especially the situation with the other office and their competing set of icons, differently than the clients. For example, although it is not entirely unusual or out of place in an organization for another project team to complete a task and present its solution to the task first, the students were deflated by this. They did not seem to realize that the technical writers also were disappointed. The writers had been vying and positioning for their own status and authority in the organization, and their colleagues had just upstaged them.

From situations such as these, researchers can compare the processes and priorities of academic and workplace contexts as they pertain to particular genres. In this case, the concerns that motivated the clients—obtaining an appealing and informative set of icons that would earn them acceptance and praise by their own supervisors—were both similar to and different from the concerns that motivated the students—developing an appealing and informative set of icons to earn praise both from the clients and from me (respectively their workplace supervisors and the "issuer" of course grades). The technical writers were concerned about how their own performance would be evaluated, just as the students were. Performance evaluation in the workplace is similar to that in academia. These "boundary areas" are sometimes elusive: In their communications with workplace clients over product design and delivery, students and even their instructors may miss or not interpret these similarities as such. They may lack experience with or exposure to the workplace genres, subtexts, and contextual forces that condition the genres. They may not recognize, for example, when a constant positioning for organizational recognition is embodied in a communication, a dynamic that is important in industry but perhaps less so in the classroom.

Russell (1997) attempts to explain this masking of boundary areas in terms of distinct activity systems. He points out, for example, that single texts can function as different genres in the two settings. In other words, students doing the "same" writing task in a classroom may be operating out of different activity systems and, therefore, writing different genres (pp. 518–519). Thus, the same writing tasks may end up being quite different in the two cultures because the activity systems of the two worlds remain distinct. Such a finding provides an alternative way of looking at the differences that may separate the cultures. The different activity systems, for example, are structured by different notions of how much time is feasible for completing various tasks, and by different notions of what may be involved in carrying out those tasks.

Russell's point here is important. He suggests that the goal is not necessarily to seek a one-to-one correspondence between the cultures. Instead, it is to research and acknowledge their complexity, and to read what is occurring in them, and, by doing so, to develop a greater awareness of how the subtexts and markers of genres in the two contexts (the workplace and the academy) are both alike and different. Those who teach technical communication need to better understand and convey these similarities and differences to our students. The key here is to see past the "product" (the quality of the output) to the process. A large amount of the "end result" is about getting there: The product "contains" all of the tracings of joint effort, competition, resource scarcity, and the like, but the situation that occurred during the icon project illustrates the need to also attend to process. In this situation, the students did not read the genre the same way the clients read it. Instead, they thought about what happened in terms of not doing well and not being appreciated for their work. They were more attuned to the product and the deliverables than to what really was going on in the situation — the consequences and implications of social choices on how decisions are enacted and resources are allocated in organizations. Had they read the situation alternately, they might have responded and acted much differently. The students misrecognized the genre as "you're just students and you didn't meet the grade." If they had recognized the genre with all of its organizational trappings, they may have seen it more for what it was or should have been: "Let's be hopeful and keep fighting." Instead of reacting dejectedly and just going through the motions of finishing the project for the course grade, they might have started thinking competitively and focused on helping the clients regain their footing.

Much more is involved here than simply process and product. Competition and other social and political dynamics play out and affect the product. These are subtleties of genres that are not always explored and acknowledged, especially in classrooms, but clearly they have an impact on communications. Communications rest on relationships of cooperation, competition, power, and numerous other factors that students, who tend to focus more on the product, do not usually discuss and often miss. By seeking a common ground, we can perhaps better understand these subtleties that get built into genres and how they get built in.

What occurred here underscores the importance and need to research, recognize, and teach the social dimensions of genres. Part of what we need to uncover and understand better through our research are these beneath-the-surface kinds of issues — status and authority, criteria and priorities for decision making, standards of work processes, and work flows.

Classroom–workplace projects, despite some elements of artificiality that may be inevitable, can help researchers identify and better understand differences in genre markers and how to deal with them. Examining project activities and outcomes, especially in relation to those projects carried out exclusively in the workplace, may reveal boundary spaces, important areas of overlap between the two cultures that are emergent and generative parts of a unity. Academics and students alike have not sufficiently practiced working in these boundary spaces partially because they focus on product more than on process, and they have yet to recognize that the product is a cumulative result of *all* the communications and material constraints that lead up to it.

ADDITIONAL BENEFITS OF RESEARCHING
A SHARED GROUND

Focusing on classroom–industry projects offers benefits beyond uniting technical communication academics and practitioners. Such projects also can help our
students to become reflective practitioners and provide them with research skills
needed for effective communication and for achieving common goals.

By analyzing the workings and effects of language, genre, and rhetorical frameworks in cooperative workplace projects, students become primed to become
reflective practitioners or practitioner researchers in their own careers. In developing a case for a sociotechnological agenda in nonacademic writing, Duin and
Hansen (1996) argue for practitioner research: "Nonacademic writers have yet
to be empowered with the skills to study themselves from the inside. Only when
nonacademic writers gain the skills to transform themselves from the inside out
will relevance in research and curriculum be achieved" (p. 13). They comment
further on the role of academics in helping students acquire these skills: "Nonacademic writing instructors must equip writers with anthropological, social
science, and linguistic skills (e.g., participant observation, journal keeping, interviews, analyses of electronic messages) that will enable them to analyze their
sociotechnological writing environments as well as participate in them" (p. 13).
Writers who possess these research skills can effect change without having to
depend on either academia or the professional site. Also, once they become able,
on their own, to describe and evaluate the genres of oral and textual communication that typify the professional world, they can train and assist their colleagues to do the same.[4]

I observed this shift from student to practitioner researcher in some graduate
students who participated in classroom–workplace collaborations and then
ended up working for their clients as interns and even permanent employees.
For example, one student currently is evaluating the effectiveness of using various types of multimedia in presenting online help information. This student
had worked on an online help file on extreme sports for a client who wished
to use this file as a demo on her company's Web site. My student decided to edit

---

[4]Two concepts in Giddens' (1984) theory of structuration also are relevant to this sort of project: the knowledgability of agents and the reflexive monitoring of social conduct. Giddens defines
reflexivity as the monitored character of the ongoing flow of social life: "To be a human being is to
be a purposive agent, who both has reasons for his or her activities and is able, if asked, to elaborate
discursively upon those reasons" (p. 3). All of this, for Giddens, constitutes a theory of action that
recognizes human beings (new practitioners in this case) as knowledgeable agents (p. 30). Giddens'
work, more generally, emphasizes the roles of individual actors in bringing about change at both
local and more global levels. Giddens says one of the most important aims of social research, therefore, is the investigation of what agents already know; the articulation of the knowledge they possess and apply in their day-to-day actions and encounters (pp. 22, 328). Practitioner-researchers
are particularly well situated to carry out this kind of inquiry.

and complete this project for the client and is now experimenting with incorporating multimedia into the file, which she plans to test for its effectiveness. This student's research will supplement the client's knowledge of the usefulness of multimedia in online help while also contributing new knowledge to our field. By integrating academic concerns with practitioner ones, this student is demonstrating through her work how the two cultures may unite in a productive dialogue that enhances knowledge and practice.

This, and other similar work can have significant implications for connecting academic and workplace cultures. This work also can bring academic ideas in line with everyday workplace practice in ways that research that is exclusively academic may not. Further, because it bridges the two settings, this work can also shed additional light on the subtle and complex features of the genres used in the two cultures. We thus should encourage and provide students with skills to engage in professional research projects, and, in doing so, create further connections and conversations between our two worlds.

LOOKING AHEAD

Paré (chapter 4, this volume) examines cultural differences between academia and the workplace and attributes to these differences much of the problem of our failure to interact effectively. He shows how the differences are ideological and epistemological, and how they appear in the values, knowledge, literacies, and activities that occur in each setting. He concludes that the failure of academics to have an impact on the workplace is tied to our underestimating these differences. In this chapter I argue that our failure also is tied to our underestimating the similarities and points of overlap between the two worlds and their genres and our need to cultivate this area of overlap. I propose a larger framework for bringing the two cultures together, grounded in research addressing their overlapping aspects. My work emphasizes examining the boundary space or middle ground between the two contexts and identifying both shared and divergent features of their genres and cultures. This focus requires delving into communication structures and genres of the two domains and all of the subtexts, social circumstances, and motivations that underlie them. This direction will help us understand better the differences and similarities between the two worlds and to develop more productive strategies for communicating across them.

The goals that we can achieve by pursuing directions such as these in our research are significant. They include, but are not limited to, bringing the concerns of the two worlds closer together, improving communication between them and developing a shared language, and doing more to cultivate shared goals and values. They also include having an impact on new information technologies and how we use them to communicate, and working out a collaborative and shared identity toward this end as the field changes and becomes more complex. Overall, these goals will help move us forward as we construct (and continually reconstruct) our identities and roles in the larger multidisciplinary milieu in which we are increasingly functioning.

ACKNOWLEDGMENT

I would like to thank Barbara Mirel for her insightful comments and feedback on my drafts of this chapter.

REFERENCES

Anson, C. M., & Forsberg, L. L. (1990). Moving beyond the academic community: Transitional stages in professional writing. *Written Communication, 7,* 200–231.

Applebee, A. (1987). Musings. *Research in the teaching of English, 21,* 5–7.

Berkenkotter, C., & Huckin, T. N. (1995). *Genre knowledge in disciplinary communication: Cognition/culture/power.* Hillsdale, NJ: Lawrence Erlbaum Associates.

Blakeslee, A. M. (1999a). *Bridging the workplace and the academy in rhetorical education: Teaching professional genres through classroom–workplace collaborations.* Paper presented at The Penn State Conference on Rhetoric and Composition, University Park, PA.

Blakeslee, A. M. (1999b). *Teaching discursive practices: Using classroom–workplace collaborations to help students learn professional genres.* Paper presented at the Association of Teachers of Technical Writing Second Annual Meeting, Atlanta, GA.

Blakeslee, A. M. (2001). Bridging the workplace and the academy: Teaching professional genres through classroom–workplace collaborations. *Technical Communication Quarterly, 10,* 169–192.

Burnett, R. E. (1996). "Some people weren't able to contribute anything but their technical knowledge": The anatomy of a dysfunctional team. In A. H. Duin & C. J. Hansen (Eds.), *Nonacademic writing: Social theory and technology* (pp. 123–156). Mahwah, NJ: Lawrence Erlbaum Associates.

Cochran-Smith, M., & Lytle, S. L. (1999). The teacher research movement: A decade later. *Educational Researcher, 28,* 15–25.

Duin, A. H., & Hansen, C. J.. (1996). Setting a sociotechnological agenda in nonacademic writing. In A. H. Duin & C. J. Hansen (Eds.), *Nonacademic writing: Social theory and technology* (pp. 1–15). Mahwah, NJ: Lawrence Erlbaum Associates.

Fleischer, C. (1994). Researching teacher-research: A practitioner's retrospective. *English Education, 26,* 86–124.

Freedman, A., Adam, C., & Smart, G. (1994). Wearing suits to class: Simulating genres and simulations as genres. *Written Communication, 11,* 193–226.

Freedman, A., & Medway, P. (1994). New views of genre and their implications for education. In A. Freedman & P. Medway (Eds.), *Learning and teaching genre* (pp. 1–22). Portsmouth, NH: Boynton/Cook Heinemann.

Giddens, A. (1984). *The constitution of society: Outline of the theory of structuration.* Berkeley, CA: University of California Press.

Hillocks, G. (1986). *Research on written composition.* Urbana, IL: National Council of Teachers of English.

Knoblauch, C. H. (1989). The teaching and practice of "professional writing." In M. Kogen (Ed.), *Writing in the business professions* (pp. 246–266). Urbana, IL: National Council of Teachers of English.

Luke, A. (1994). Series editor preface. In A. Freedman & P. Medway (Eds.), *Learning and teaching genre* (pp. vii–xi). Portsmouth, NH: Boynton/Cook Heinemann.

Moss, B. (1992). Ethnography and composition: Studying language at home. In G. Kirsch & P. A. Sullivan (Eds.), *Methods and methodology in composition research* (pp. 153–171). Carbondale: Southern Illinois University Press.

North, S. M. (1987). *The making of knowledge in composition.* Upper Montclair, NJ: Boynton/Cook.

Ray, R. E. (1992). Composition from the teacher-research point of view. In G. Kirsch & P. A. Sullivan (Eds.), *Methods and methodology in composition research* (pp. 172–189). Carbondale: Southern Illinois University Press.

Ray, R. E. (1996). Afterword: Ethics and representation in teacher research. In P. Mortensen & G. E. Kirsch (Eds.), *Ethics and representation in qualitative studies of literacy* (pp. 287–300). Urbana, IL: National Council of Teachers of English.

Russell, D. R. (1997). Rethinking genre in school and society: An activity theory analysis. *Written Communication, 14,* 504–554.

Spinuzzi, C. (1996). Pseudotransactionality, activity theory, and professional writing instruction. *Technical Communication Quarterly, 5,* 295–308.

Wickliff, G. A. (1997). Assessing the value of client-based group projects in an introductory technical communication course. *Journal of Business and Technical Communication, 11,* 170–191.

Winsor, D. A. (1996a). Writing well as a form of social knowledge. In A. H. Duin & C. J. Hansen (Eds.), *Nonacademic writing: Social theory and technology* (pp. 157–172). Mahwah, NJ: Lawrence Erlbaum Associates.

Winsor, D. A. (1996b). *Writing like an engineer: A rhetorical education.* Mahwah, NJ: Lawrence Erlbaum Associates.

4

Keeping Writing in Its Place: A Participatory Action Approach to Workplace Communication

ANTHONY PARÉ
McGill University

For the past 15 years, I have moved back and forth across the border between academia and the workplace as a writing teacher, researcher, consultant, trainer, and advisor. Although my full-time position is within a university, I have sought to influence workplace literacy practices and policies by going into the field. This chapter explains some of the difficulties I have faced in my work as a university-based literacy consultant and describes some of the strategies I have used to overcome those difficulties. I believe the difficulties help explain the tensions and misunderstandings that exist between academia and the workplace, and I hope the strategies might be useful to those who seek to change professional literacy practices.

My specific disciplinary interest has been social work and its allied fields and, as a result, I have conducted courses, workshops, seminars, and research studies in agencies, hospitals, group homes, neighborhood clinics, community centers, and schools. In addition, I have spoken at the annual meetings of national social work associations and in university schools of social work across Canada. In all that time, I have worked with thousands of social work students, educators, and practitioners, and yet I can claim only modest success in my attempts to use my knowledge of composition theory and my familiarity with social work discourse to effect change in professional literacy practices. I believe the limitations of my success are due to the inescapably social nature of writing—its deep and complex implication in human activity—and in the pages that follow I hope to elucidate that simple, essential fact.

In 1985, when I decided to study writing in nonacademic settings for my dissertation (Paré, 1991a), I was inspired by the pioneering studies of people such as Knoblauch (1980), Bataille (1982), Faigley and Miller (1982), Odell and Goswami (1982), and Selzer (1983). For me and many others, these scholars had begun to free writing from the individual mind, where cognitive theories had confined it, and to relocate it in the world. We were eager to follow it out there. In the year I began my doctoral studies, my interest in workplace writing was much supported and confirmed by the publication of Odell and Goswami's *Writing in Nonacademic Settings* (1985), and in the year I finished doctoral work, Bazerman and Paradis' (1991) important collection of essays appeared. Soon after graduation, I was able to contribute to this growing field of scholarship with a chapter in Spilka's edited collection, *Writing in the Workplace: New Research Perspectives* (1993; see also Paré, 1991b).

Since those early days, I have continued to study workplace writing, which Cooper (1996) calls "the most exciting area of research and scholarship in writing" (p. x). And, yet, that exciting work has not had the impact one might expect or hope. Moreover, as the chapters in this book indicate, the failure to influence has been bidirectional: composition theory and practice developed in academia have had only a limited effect in the workplace, and the insights gained through studies of workplace writing have not much changed composition instruction in the schools. Why?

THE EMBEDDEDNESS OF WRITING

My own initial forays into professional social work settings might offer the beginnings of an answer to that question. In 1985 and 1986, I studied a team of social workers attached to Quebec's Youth Court. Their clients were adolescents who had run afoul of the law, and their chief writing task consisted of reports on those youths—reports that could be read by other social workers, judges, lawyers, police officers, doctors, psychiatrists, psychologists, and school principals, as well as by the client and the client's family. I observed and interviewed the workers, collected composing-aloud protocols from them as they wrote, read dozens of their reports in draft and final copy, and tested my interpretations of their activities in postanalysis interviews with the workers. Two observations on that experience seem worth sharing.

First, when I asked about their "writing," the social workers looked embarrassed, disclaimed any ability to write, and lamented their inattention to their high school English teachers.[1] Moreover, they claimed to do very little writing,

---

[1] This response is familiar to anyone who has been introduced at a cocktail or dinner party as an "English" teacher. At such times, people frequently express concern about their language ability, particularly their supposedly poor grammar. I believe the response is an indictment of our ap-

despite the fact that I could see that they spent many hours a week on the reports required of them. After a while, I realized that they referred to that activity as "recording," not "writing." That is, they named the purpose or object of the activity, not the way in which the activity was realized. What they were *doing* was recording, or making a record, and writing was how they did it.[2] In Polanyi's (1964, 1967) terms, the recording was in focal awareness, the writing in subsidiary awareness; writing was what they did in order to get something else accomplished.

This role of writing in the service of something else is a key part of the reason that it is difficult to bring insights from workplace writing research back into the classroom. No classroom-based simulation of nonacademic writing can capture the complexity or intricacy of the original rhetorical context, nor can it easily make writing subservient to an institutional or community goal beyond the act of writing itself (Dias, Freedman, Medway, & Paré, 1999; Dias & Paré, 2000). But it also helps explain why attempts to influence writing in the workplace may have only limited effect: Writing in professional contexts serves particular ends, and unless those ends are changed, writing practices will remain the same. As a result, when literacy workers—those from universities or those with workplace positions—seek to change texts or textual practices, they must focus on, and expect, systemic change. Writing is not separate from the actions and activities it serves.

Second, I became aware as soon as I entered the workplace that writing on the job can be dangerous. The form, content, distribution, and use of many professional texts are closely governed by both implicit and explicit guidelines and regulations, and failure to comply may place individuals in jeopardy. Workers and their ideas receive exposure when texts become public and circulate within organizations, and there are a host of potentially negative responses to any given document, ranging from mild rebukes from supervisors to serious litigation concerns. The texts produced by the social workers I observed left them open to a variety of risks: Legal action by clients' families, reprimands or fines from judges if they failed to abide by legal regulations, and public charges of bias or incompetence from defense lawyers during court hearings.

One early success I had in transporting workplace dynamics into the classroom was based on the observation that professional writing is often dangerous and socially difficult. To exploit that charged reality of workplace practice, I had social work students in a composition course write an assessment of a friend's or relative's ability to do something (e.g., succeed at university, quit smoking, establish healthy relationships, deal with crisis). I then told them that similar

---

proaches to literacy education, many of which continue to stress linguistic correctness over rhetorical consequence.

[2]A number of composition researchers have used Activity Theory to demonstrate how the *action* of writing can be involved in the performance of quite different *activities* (e.g., Dias, 2000; Dias, Freedman, Medway, & Paré, 1999; Russell, 1995, 1997).

assessment documents are accessible to clients in professional practice, and asked them to show their assessment to the person they had written about and to collect a written response from that person. Students begged me not to require the second and third steps in the assignment, and because their obvious discomfort indicated that the exercise had made its point, I agreed. I continue to use variations on this exercise with students and practitioners. Making friends or relatives the subject of written assessments raises issues of objectivity, power, and interpersonal relations more starkly than using clients, with whom even students have developed a distant, "professional" relationship.

In addition to the tensions and power struggles played out between writers and readers through textual practices, professional texts enter a highly competitive "linguistic marketplace" (Bourdieu, 1993), and social work documents often vie for attention and status with documents from more prestigious disciplines, such as law and medicine (Paré, 1993, 2000). There are multiple and serious consequences of workplace writing, and they are often difficult to predict or anticipate. That is why academics and researchers who enter the workplace must be humble and sensitive: It takes a long time to understand another's culture, and great nerve to seek to change it (Segal, Paré, Brent, & Vipond, 1998).

These two observations point to the extreme embeddedness of workplace writing. As Gee (1996) points out, when we examine the full, complex operations of any particular group, "it is next to impossible to separate anything that stands apart as literacy practice from other practices" (p. 41). In other words, the act of writing is always enmeshed in a larger social action. That is the simple explanation for the difficulties we encounter when we move between school and workplace: Since workplace writing is always inextricably connected to local, situation-specific practices, any attempts to replicate or "teach" it in the classroom are doomed to failure because the act loses meaning out of its context and is disassociated from its purpose. And because workplace writing is always regulated and usually implicated in struggles for power and status, it is perilous to tinker with or influence its practice, especially if one is an outsider, as many of us are when we work as researchers, consultants, or trainers in workplace settings.

To grasp the full implications of the social embeddedness of writing, and to fashion strategies for teaching and talking about writing as a situated activity, two contemporary perspectives on learning and language are helpful: theories of situated learning, and critical approaches to literacy.

SITUATED LEARNING

Some contemporary theories of cognition and learning emphasize the cultural nature of knowledge and knowledge-making. These new theoretical perspectives signal important shifts:

• Away from a view of knowledge as fixed, universal, and generalizable, toward a view of knowledge as shifting, dynamic, local, and relative.

• Away from an understanding of cognition as mental processes inside individual heads, toward a conception of cognition as a social, collective, and distributed activity.

• Away from a view of learning as the accumulation of discrete skills and context-free knowledge, toward a view of learning as the gradually increasing ability to participate in socially-situated, collaborative practices.

• And, finally, as a result of these shifts, a move away from the individual as the unit of attention and analysis in social science research, toward a focus on collective activity in disciplines, organizations, institutions, and other "communities of practice" (Lave, 1991).

The specific theoretical frameworks supporting studies of situated learning are derived from cultural–historical (or sociocultural) approaches to the study of human activity, cognition, and learning (e.g., Cole, 1995, 1996; Wertsch, 1991). A key move of such approaches is the shift in attention from individuals to the activities in which individuals participate within communities of practice:

> This theoretical view emphasizes the relational interdependency of agent and world, activity, meaning, cognition, learning, and knowing. It emphasizes the inherently socially negotiated quality of meaning and the interested, concerned character of the thought and actions of persons engaged in activity. . . . [T]his view also claims that learning, thinking, and knowing are relations among people engaged in activity *in, with, and arising from the socially and culturally structured world.* (Lave, 1991, p. 67, emphasis in original)

The unit of analysis that emerges from this perspective might be called, to use Lave's terms, "persons-engaged-in-activity," and learning can be considered the process of gradually improving engagement or participation in particular relations and activities.

This is the critical notion from studies of situated learning that helps explain how people learn to write on the job and why teaching about workplace writing in the classroom is difficult, or even impossible: Writing, like learning, thinking, and knowing, is a relation "among people engaged in activity *in, with, and arising from the socially and culturally structured world.*" Learning to write in the workplace is part of the gradual, collaborative, and situated transformation of neophyte into expert. People learn what they need to know, and where they need to know it, and they learn it through engagement and with guidance. Rogoff (1991) calls the process "guided participation," Freedman and Adam (2000) call it "facilitated performance" and "attenuated authentic participation" (see also Dias et al., 1999), and Lave and Wenger (1991) call it "legitimate peripheral participation." In each case, the movement toward expertise might be described as centripetal: The learner is drawn from initial, hesitant performance of minor or relatively unimportant tasks toward proficiency in the community's critical activities under the direction of more experienced individuals and in collaboration with other community members. Underlying this dynamic is Vygotsky's (1978) notion of a "zone of proximal development": "The distance between the actual developmental level as determined by independent problem solving and the level of potential development as determined through problem solving under adult guidance or in collaboration with more capable peers" (p. 86).

Although Vygotsky is referring here to children, the adult's developmental trajectory is a similar series of such zones. During initial professional practice, the neophyte takes on or is given literacy tasks just outside her or his ability and assisted in the performance of those tasks by veteran members of the community. The literate ability thus gained is local, or situation-specific, as Gee (1996) indicates:

> a way of reading a certain type of text is *only* acquired, when it is acquired in a "fluent" or "native-like" way, by one being embedded (apprenticed) as a member of a social practice wherein people not only read texts of this type in this way, but also talk about such texts in certain ways, hold certain attitudes and values about them, and socially interact over them in certain ways. Thus, one does not learn to read texts of type X in way Y unless one has experience in settings where texts of type X are read in way Y. . . . One has to be socialized into a practice to learn to read texts of type X in way Y, a practice other people have already mastered. (p. 41)

My own observations of neophytes learning workplace writing match precisely the process that Gee describes (see especially Dias et al., 1999; Dias & Paré, 2000; Paré & Szewello, 1995). The implications for those of us involved in writing instruction in school and workplace are clear: Expertise is, in part, a gradual transformation that occurs in situ, under the guidance and direction of experienced members of particular communities of practice.

An important point must be made here. Neophytes do not become expert *at* writing; they become expert at *using* writing to perform or participate in something. They learn to write texts of type X in way Y to achieve Z. Without Z, there is no way Y and no text X. As a result, I have become convinced that it is not possible to "teach" professional writing at a distance from the practices it serves and the contexts within which it operates. It may be that there are "typical" texts in circulation in various workplace settings, and drawing students' attention to those texts might well be instructional, but no amount of play-acting in school can capture the actual workplace drama from which the texts arose.

Along with colleagues and graduate students, I have been involved in long-term studies of writing in the academic and workplace settings of a number of different disciplines (Dias et al., 1999; Dias & Paré, 2000). One focus of those studies was on school-based simulations of professional writing (Freedman & Adam, 2000; Freedman, Adam, & Smart, 1994). Our conclusion concerning case-study writing and other forms of simulation is that much can be learned through such activities, but not about actual workplace writing—at least, not when students have little or no experience of work. However, instructional attention to professional writing does seem to have an impact on the actual practice during periods of field education (internships, work–study programs, apprenticeships), especially if that instruction is provided by veteran practitioners while students are attempting to produce authentic workplace texts.

CRITICAL APPROACHES TO LITERACY

Theories of situated learning allow us to consider how acts of workplace writing are deeply embedded in community activities, from which they cannot be extricated and relocated to the classroom for the purposes of writing instruction. A common notion in much contemporary theorizing, *discourse*—broadly conceived as the social activity of making meaning through language and other symbol systems—allows us to consider the situatedness of writing in somewhat different terms. When used in reference to the meaning-making practices of a particular community, the concept of discourse often encompasses typical contexts or settings, participants, roles, modes (speaking, writing, hypertext, etc.), rules, topics, forms/formats, genres, and so on. However, the term also carries connotations of ideology, and there has emerged a rich and provocative body of research into the ways in which power works in and through literacy practices. This perspective, which Gee (1996) calls *socioliteracy studies* (p. 122), draws from discourse analysis, linguistics, semiotics, and critical language awareness (e.g., Caldas-Coulthard & Coulthard, 1996; Clark & Ivani, 1997; Fairclough, 1992, 1995; Gee, 1996; Graff, 1987; Lemke, 1995; Luke, 1988; Muspratt, Luke, & Freebody, 1997; Street, 1984, 1995). Regardless of their disciplinary origins, these socioliteracy studies share certain theoretical viewpoints: They see literacy in all of its specialized disciplinary and institutional manifestations as always local and located (never universal, never transcendent), always saturated with values and beliefs (never neutral, never merely a vehicle for communication), and always shaped by and for particular social actions and ends. Furthermore, they acknowledge the formative effects of literacy—that is, the cognitive, social, cultural, and ideological influence of literacies on those who participate in them.

This social perspective has revealed the extent to which individuals are influenced by collective literacy practices, a phenomenon captured by Gee's (1996) reference to discourses as "identity kits." Gee explains: "Discourses are ways of being in the world, or forms of life which integrate words, acts, values, beliefs, attitudes, and social identities, as well as gestures, glances, body positions, and clothes" (p. 127). Fairclough (1995) makes a similar claim when he argues that "in the process of acquiring the ways of talking which are normatively associated with a [particular] subject position, one necessarily acquires also its ways of seeing, or ideological norms" (p. 39). Herndl (1996a) calls this "the ideologically coercive effects of institutional and professional discourse," or, "the dark side of the force" (p. 455; see also Herndl, 1996b). Of course, a particular discourse need not be "dark," but all discourses are undeniably coercive; that is, they influence one's worldview. Furthermore, and critical to the focus of this book, is the fact that all participants in a discourse are located in a web of power relations that are played out in and through literacy.[3]

It is essential for those of us teaching in or studying the workplace to understand this: Power struggles are not merely reflected in institutional discourse, they are actually waged through discourse. When we encourage individual workers to alter their writing processes or products, when we advise organizations to develop new documentary procedures, when we suggest the redesign of forms, when we seek to "improve" a community's literacy practices, we are tampering with something fundamental and dynamic. Below the surface of style, format, and procedure move the values, beliefs, and attitudes that shape the community, and even apparently minor adjustments can have profound effects.

Some of the most common writing-course exhortations could be disastrous if followed in the workplace: avoid the passive voice, write clearly, state your topic, be explicit and direct, address your reader, make your transitions obvious, reach a firm conclusion, support your opinion with facts. This advice, reasonable perhaps within the confines of the school, where the student works and is evaluated as an individual, might well cause serious problems for the worker operating as a member of a community. Many workplace writers disappear into the passive voice because it offers them protection; overly direct requests or rebuffs can backfire; the actual primary reader of a document may be listed, as if unimportant, on the distribution list; and all the facts in the world may not support an opinion counter to that held by one's boss or by a member of an overlapping but more powerful community of practice.

This last point is especially important. Many social workers must tailor their documents for the use of doctors, lawyers, and other higher status professionals (Paré, 2000). Failure to do so might jeopardize their position within the institution. For example, if doctors found social work reports of no value, would hospitals continue to employ social workers? As a result, hospital-based social workers I have met often make their reports short so doctors will read them, despite the fact that short reports frequently lack the detail and complexity required in effective social work.

---

[3]Some genre studies have described this ideological tension within institutional discourse (e.g., Coe, Lingard, & Teslenko, in press; Paré, in press; Schryer, 1994). In fact, although they differ in some significant ways (Freedman, 1993), the Australian school of genre studies (e.g., Cope & Kalantzis, 1993; Reid, 1987) and its North American counterpart (e.g., Freedman & Medway, 1994a, 1994b; Miller, 1984) share many theoretical perspectives with scholars working in what Gee (1996) calls "socioliteracy studies."

TOWARD SOME SOLUTIONS:
A PARTICIPATORY ACTION APPROACH

To acknowledge the embeddedness of workplace writing—indeed, to exploit it —I have found it useful to draw on some of the tenets of participatory action research (e.g., Park, Brydon-Miller, Hall, & Jackson, 1993). According to Tandon (1988), participatory action research "is based on the belief that ordinary people are capable of understanding and transforming their reality" (p. 13). Tandon argues that such research must allow participants to set the research agenda, participate in data collection and analysis, and exercise control over the whole research process, including the use of outcomes (p. 13). Ideally, the distinction between researcher/trainer and participant disappears in this type of work, and for that to happen a first step is critical: The problem to be addressed (or question to be answered) must come from the community members. The writing "expert" need not be passive, of course, but must be responsive to the participants' identification of concerns, definition of problems, and statements of need. Indeed, what I have discovered is that my experience as a writing teacher and researcher has proven most valuable when applied to a concern selected by those with whom I am working.

As a result of adopting a participatory stance, the difference between my work as a researcher and my work as a workplace consultant and trainer has gradually diminished over the past few years, and I now see any opportunity to enter the workplace or to work with professionals as a chance to study nonacademic literacy and to take part in efforts to make literacy practices more effective, more responsive to the needs of community members, more transparent, and less inflexible.

My recent (and still modest) success contrasts with years of some frustration in my work with practicing social workers. It was not that workers were unresponsive to my presentations or workshops; on the contrary, composition studies provided me with a persuasive theory and highly effective pedagogy, and my research allowed me to get an insider's view of social work literacy. As a result, I was able to describe social work writing conditions with an accuracy that workers often found uncanny, and my workshop techniques generally elicited enthusiasm. People thought I was a social worker. But—and this is my point— *nothing changed as a result of my work.* Participants would leave my sessions determined to apply the lessons learned, but they were my lessons, my solutions, my insights. Back in their offices, the crush of cases, the force of habit and institutional history, and the seemingly inexorable activity of social work practice pushed new ideas about recording into the background. Weeks later, I would discover that the plans to revise recording practices that were hatched during a workshop had quietly died. The goals those practices served had not been changed, and neither had the culture that shaped those practices.

The reason my interventions had such limited and temporary success became clear only after I began to work with Inuit social workers from arctic Quebec. Although I felt I knew plenty about social work writing, I was at least humble enough to admit that I knew next to nothing about the Inuit culture and how social work operated in far northern communities. I decided to let the workers tell me about the practice and problems of social work and, more specifically, about the difficulties they experienced with recording. Although I did not know it at the time, I was taking a participatory research stance.

"White People Are Greedy for Other People's Problems"

McGill University helps to run a Certificate in Northern Social Work Practice, and I have been privileged to work with students in that program, all of whom are Inuit women and men working as social workers in small communities on Quebec's northern coast. The extreme isolation of those communities meant that the full impact of mainstream North American culture did not hit until relatively recently, but when it did, it hit hard. As they struggle in the transition between their traditional ways and the new life forced on them from the south, the Inuit people are suffering cruelly high per capita percentages of suicide, substance abuse, and physical and sexual violence. During my first meeting with these extraordinary people—who seek to reconcile traditional community-healing practices with the professional social work imported from the south—one worker was blunt in her assessment of professional recording practices: "White people are greedy for other people's problems." Another said that record-keeping was like "stealing someone's life."

As the workers spoke about their difficulties with recording, a central problem became clear: Participation in conventional social work literacy pushed the workers toward the detached professional self that was essential in the south, where workers and clients live apart and have no relationship outside the interview, the office, or the courtroom. "The text," says de Montigny (1995), "is a mask concealing the embodied speaker who utters this or that claim. Through the text, social workers can promote their claims as though these were the universal wisdom of the profession in general" (p. 64). But the Inuit workers live and work in their communities, and their clients are their neighbors, friends, and families. Transporting textual practices to the north meant transporting, as well, the elements of context and culture that had created and sustained them: the impersonal, detached persona of professional life; the anticipated narratives of southern social work clients; the categories, lifestyles, values, beliefs, and power relations of the urban welfare state. As a result, the Inuit workers were forced into a role ill-suited to their actual living and working situations.

On the other hand, failure to participate in professional literacy practices made the Inuit workers outcasts in their professional community, where their managers and supervisors are, for the most part, nonaboriginal, university-trained social workers. One worker explained the dilemma this way:

> I have to satisfy both distinct cultures: the paper-work culture [White bureaucracy] . . . and my culture, you know, who I am, who [my clients] are, the way we speak, the way we talk. If I become a professional person with my family, I'm not going to have any more family . . . I'm going to push them away."

In planning for a recent course called Culture and Communication, my coteacher, Laura Mastronardi, and I decided to take our lead from the Inuit workers. Although we talked for hours, collected materials, and considered exercises and assignments, we did not have the slightest idea what to do on the first morning of the course: no lesson plan, no outline, no schedule, no readings, no evaluation procedures, no assignments, no deadlines. We did have a belief, which we shared with the students, that they were the experts on the topic of culture and communication; they were the people who day in and day out had to explain their Inuit culture to their southern, White managers, and their social work culture to their Inuit neighbors, families, and friends; they were the people who knew what the problems were and what the possible solutions and strategies were. We had materials, a few ideas, and a willingness to help where we were needed. We indicated our areas of specialization: Laura's years of work as a professional social worker; my own experience as a writing teacher and researcher. But we ceded authority to them; we acknowledged their expertise. The result of that move, which they clearly welcomed, was dramatic.

For the first day and a half, they brainstormed and defined problems: What did they find difficult in their daily speaking, writing, reading, and listening? Why did they have difficulty communicating with supervisors? Clients? Fellow workers? A complex picture of multiple identity emerged, of ambiguous boundaries between personal and professional lives. For years, I have attempted to convince southern, White social workers to drop the professional mask, to write in the active voice, to speak to and about their clients as people. But in the north, such a division between the personal and professional is impossible: the Inuit workers walk out of their offices into the communities they serve. Their clients live next door, play with their children, are related. I had to learn this. I had to learn its rhetorical implications.

We began to shape a course around their needs and concerns. What was most profound about the experience was that we erased the distinction between teacher and student. Or, rather, we moved back and forth between those roles as our needs and our experience allowed. At any moment, on any given topic, someone would come forward as teacher. I believe that by rejecting the role of experts, Laura and I made it possible for all of us as a group to negotiate a space between cultures, a space where teaching and learning were the natural outcomes of a common and collective need to know, and where the roles of teacher and student were constantly interchanged. In the process, we were all transformed.

Typical of the work produced during the first days of the course are the sentences listed in Appendix A. As workers described recording situations and the actual people involved in the situations, it was possible to generate specific rhetorical strategies to respond to the problems they identified. The students felt that some of their readers lacked respect, both for the writers themselves and the Inuit community in general, and we created phrases that helped establish the writers' authority or that sought to build mutual respect. They felt awkward stating their opinions, but agreed that they wanted their views known, and so we devised rhetorical ways to advance opinion without being aggressive. In many cases, once the problem had been defined, workers would offer each other solutions.

However, as one might expect, we began to see how deeply enmeshed these communication problems were with other aspects of social work, and how the cross-cultural gaps—between Inuit and non-Inuit, and between social work and the community—created tensions and suspicions. In the last days of the course, the workers decided to prepare a report that could be circulated in the northern social work community in order to influence policy and practice. The result was an extraordinary, 2-day session of group authorship that produced a seven-page report, an excerpt of which is included as Appendix B.

Perhaps it has become cliché to talk about empowerment, but the Inuit workers definitely left the course feeling more powerful than they did when they began, and neither Laura nor I had given them that power; they had created it themselves. Certainly, we had helped to establish a context within which such work was possible, but they had been the ones to examine closely the complex social conditions in which they live and work, and they had fashioned strategies to improve their own and others' literacy practices. My experience with these workers has affected all of my workplace practice, even my 1-day workshops with workers in more conventional social work settings.

BIG PROBLEMS, LONG-TERM SOLUTIONS

It is no exaggeration to say that there is a crisis in social work recording. At one time, social workers were expected, even encouraged, to keep detailed, speculative records of their clients' lives. It was acknowledged that the process of articulating the complexity and intricacy of an individual's situation would lead the recording worker into the critical thinking necessary for the professional practice of social work. In addition, extensive records served both individual and institutional memory, so that the fullness of a client's life could be recovered on file.

However, for a variety of reasons and from a variety of perspectives, the lengthy and thorough records of the past are no longer considered desirable: They take too much time to write; they take too much time to read; they interfere with the clients' right to privacy; they put workers at risk of litigation; they do not serve the needs of doctors, lawyers, psychologists and other allied professionals; they serve the needs of bureaucrats only, and do not assist workers; they are too subjective and judgmental. The list goes on. And yet, recording is still required and necessary for the same reasons it has always been: It has the potential to engage workers in critical thinking, and it creates a permanent account.

This tension, which places workers in an untenable situation between the need to record and the many reasons against recording, has been the focus of all my workshops over the past few years. The process I follow is much the same from session to session: Participating workers form small groups, conduct a critical analysis of the problems associated with recording, and work together to find solutions.[4] Each group creates a written list of the problems and solutions they have identified, and that list is presented at a plenary session. If time allows, lists are revised. Finally, all written material is submitted to me, and, either alone or with willing workers, I prepare documentation. At times those documents have been checklists or guidelines for recording practice within an agency or hospital, at other times detailed reports have been prepared and submitted to documentation committees, managers, supervisors, and other administrators.

Appendix C consists of a section of such a report. The material for the report, which is 15 pages long, was generated during 2 days of workshops with approximately 100 workers associated with a Canadian provincial Ministry (i.e., government department) of social services. The report contains their explanation of the difficulties with recording and their recommended solutions to those difficulties. The excerpt in Appendix C concerns one of the most vexing problems for workers: the demand for "objectivity" in social work recording. Through a close and critical examination of the concept of "objectivity" as it is used in social work discourse, workers were able to locate themselves in the web of tensions surrounding every record. On the one hand, the possibility that clients might read their records, or that workers might end up in court, demanded a certain level of precision—the kind of bias-free observation thought to be associated with science. Such "objectivity" also allowed workers to keep records short and to avoid reprimands from superiors or allied professionals. On the other hand, brief records were of little value in on-going work with clients, or when clients were transferred from one locale to another. Moreover, increasing accountability has meant that workers can get in trouble for failing to record pertinent details—a threat reduced by including plenty of detail.

---

[4]Sometimes I use a series of questions to help the workers start their analysis, especially if time is short: Why do you produce records? Who reads them? Why do they read them? What do you find difficult about recording? How would you like to change recording formats and procedures? And so on. Workers never need encouragement to vent their frustration with recording, but they often stray from the topic; the embeddedness of writing means that any workplace discussion of it quickly leads to many other issues.

The workers recognized that this and other aspects of recording implicated them in institutional power relations, and asked that the report be disseminated as far as possible and used as a blueprint for long-term change within the Ministry. Each workshop participant received a copy of the report to be shared with coworkers, supervisors, and managers, and copies were sent to key people throughout the Ministry. The goal here, as in the course with the Inuit workers, was to effect change within the workplace culture, so that any change to literacy practices would be lasting.

CONCLUSION

Because I have acknowledged only modest success, it seems presumptuous to offer advice, particularly because every profession and every workplace context is unique. However, I believe there are some general guidelines that can help academics improve their chances of influencing workplace literacy.

First, academics need to allow the definition of problems to arise from workers. Academics might be able to convince workers that their definitions of their literacy problem are accurate, but they will then solve their own problem, not that of the workers. To take this approach, academics must be patient, ready to learn, and willing to spend the time and effort it takes to immerse themselves in another culture. The quick literacy fix imposed by many consultants and workplace trainers usually comes undone soon after their departure, but the results of a participatory action approach can have a lasting effect.

Second, academics must be aware of and sensitive to whatever it is that writing *does* in the workplace. Literacy practices serve community goals, and without an understanding of those goals, no change is possible. Adjusting a procedure or a format may seem relatively easy in a writing workshop or seminar, but such changes can have widespread effects on an organization's activities.

Third, academics need to beware of the ways in which their interventions upset power relations, expose individual workers, affect readers, and otherwise influence communities within which they work. Workplace communications are fraught with implications for relationships within institutional hierarchies, and the failure to understand those dynamics reduces credibility and effectiveness.

Finally, where possible, academics should create situations that facilitate interaction among community members, so that veterans may instruct neophytes in ways that academics cannot. Although academics may not be able to "teach" workplace writing directly, they can hasten a newcomer's development by engineering contexts in which their literacy learning is supported by interaction with knowledgeable veterans. For example, in workshops, I occasionally introduce the same report written in quite different ways and ask workers to evaluate the documents. In the resulting discussion, veterans make explicit some of the (often tacit) values underlying workplace literacy practices and some of the less obvious consequences arising from texts. Similarly, setting newcomers and veterans to work on the redesign of a workplace text can lead to some rich moments of teaching and learning.

Workplace communications are complex and deeply embedded in context and culture. Those who are interested in studying and influencing professional literacy practices must be patient and humble. We can understand those practices and effect change in them, but not without respecting their profoundly social nature.

REFERENCES

Bataille, R. R. (1982). Writing in the world of work: What our graduates report. *College Composition and Communication, 33,* 276–282.

Bazerman, C., & Paradis, J. (Eds.). (1991). *Textual dynamics of the professions: Historical and contemporary studies of writing in professional communities.* Madison: University of Wisconsin Press.

Bourdieu, P. (1993). *Sociology in question* (R. Nice, Trans.). London: Sage Publications. (Original work published 1984)

Caldas-Coulthard, C. R., & Coulthard, M. (Eds.). (1996). *Texts and practices: Readings in critical discourse analysis.* London: Routledge.

Clark, R., & Ivani, R. (1997). *The politics of writing.* London: Routledge.

Coe, R., Lingard, L., & Teslenko, T. (Eds.). (in press). *The rhetoric and ideology of genre: Strategies for stability and change.* Cresskill, NJ: Hampton Press.

Cole, M. (1995). Socio-cultural-historical psychology: Some general remarks and a proposal for a new kind of cultural-genetic methodology. In J. V. Wertsch, P. del Rio, & A. Alvarez (Eds.), *Sociocultural studies of mind* (pp. 187–214). Cambridge: Cambridge University Press.

Cole, M. (1996). *Cultural psychology: A once and future discipline.* Cambridge, MA: The Belknap Press of Harvard University Press.

Cooper, M. M. (1996). Foreword. In A. Duin & C. Hansen (Eds.), *Nonacademic writing: Social theory and technology* (pp. ix–xii). Mahwah, NJ: Lawrence Erlbaum Associates.

Cope, B., & Kalantzis, M. (1993). *The powers of literacy: A genre approach to teaching writing.* London: Falmer Press.

de Montigny, G. (1995). *Social working: An ethnography of front line practice.* Toronto, ON: University of Toronto Press.

Dias, P. X. (2000). Writing classrooms as activity systems. In P. X. Dias & A. Paré (Eds.), *Transitions: Writing in academic and workplace settings* (pp. 11–29). Cresskill, NJ: Hampton Press.

Dias, P. X., Freedman, A., Medway, P., & Paré, A. (1999). *Worlds apart: Writing and acting in academic and workplace contexts* (pp. 11–29). Mahwah, NJ: Lawrence Erlbaum Associates.

Dias, P., & Paré, A. (Eds.). (2000). *Transitions: Writing in academic and workplace settings.* Cresskill, NJ: Hampton Press.

Faigley, L., & Miller, T. (1982). What we learn from writing on the job. *College English, 44,* 557–569.

Fairclough, N. (1992). *Discourse and social change.* Cambridge, UK: Polity Press.

Fairclough, N. (1995). *Critical discourse analysis: The critical study of language.* London: Longman.

Freedman, A. (1993). Show and tell? The role of explicit teaching in learning new genres. *Research in the Teaching of English, 27,* 222–251.

Freedman, A., & Adam, C. (2000). Write where you are: How do we situate learning to write? In P. X. Dias & A. Paré (Eds.), *Transitions: From academic to workplace settings* (pp. 31–60). Cresskill, NJ: Hampton.

Freedman, A., Adam, C., & Smart, G. (1994). Wearing suits to class: Simulating genres and simulations as genre. *Written Communication, 11,* 192–226.

Freedman, A., & Medway, P. (Eds.). (1994a). *Genre and the new rhetoric.* London: Taylor and Francis.

Freedman, A., & Medway, P. (Eds.) (1994b). *Learning and teaching genre.* Portsmouth, NH: Boynton/Cook Heinemann,

Gee, J. P. (1996). *Social linguistics and literacies: Ideology in discourses* (2nd ed.). London: Falmer.

Graff, H. J. (1987). *The legacies of literacy: Continuities and contradictions in Western culture and society.* Bloomington, IN: University of Indiana Press.

Herndl, C. G. (1996a). Tactics and the quotidian: Resistance and professional discourse. *Journal of Advanced Composition, 16*(3), 455–470.

Herndl, C. G. (1996b). The transformation of critical ethnography into pedagogy, or the vicissitudes of traveling theory. In A. Duin & C. Hansen (Eds.), *Nonacademic writing: Social theory and technology* (pp. 17–34). Mahwah, NJ: Lawrence Erlbaum Associates.

Knoblauch, C. H. (1980). Intentionality in the writing process. *College Composition and Communication, 31,* 153–159.

Lave, J. (1991). Situated learning in communities of practice. In L. B. Resnick, J. M. Levine, & S. D. Teasley (Eds.), *Perspectives on socially shared cognition* (pp. 63–82). Washington, DC: American Psychological Association.

Lave J., & Wenger, E. (1991). *Situated learning: Legitimate peripheral participation.* Cambridge: Cambridge University Press.

Lemke, J. L. (1995). *Textual politics: Discourse and social dynamics.* London: Taylor and Francis.

Luke, A. (1988). *Literacy, textbooks and ideology.* London: Falmer Press.

Miller, C. R. (1984). Genre as social action. *Quarterly Journal of Speech, 70,* 151–167.

Muspratt, S., Luke, A., & Freebody, P. (Eds.). (1997). *Constructing critical literacies: Teaching and learning textual practice.* Cresskill, NJ: Hampton Press.

Odell, L., & Goswami, D. (1982). Writing in a non-academic setting. *Research in the Teaching of English, 16,* 201–223.

Odell, L., & Goswami, D. (Eds.). (1985). *Writing in nonacademic settings.* New York: Guilford.

Paré, A. (1991a). *Writing in social work: A case study of a discourse community.* Unpublished doctoral dissertation, McGill University, Montreal.

Paré, A. (1991b). Ushering 'audience' out: From oration to conversation. *Textual Studies in Canada, 1,* 45–64.

Paré, A. (1993). Discourse regulations and the production of knowledge. In R. Spilka (Ed.), *Writing in the workplace: New research perspectives* (pp. 111–123). Carbondale, IL: Southern Illinois University Press.

Paré, A. (2000). Writing as a way into social work: Genre sets, genre systems, and distributed cognition. In P. X. Dias & A. Paré (Eds.), *Transitions: Writing in academic and workplace settings* (pp. 145–166). Cresskill, NJ: Hampton Press.

Paré, A. (in press). Genre and identity: Individuals, institutions, and ideology. In R. Coe, L. Lingard, & T. Teslenko (Eds.), *The rhetoric and ideology of genre: Strategies for stability and change.* Cresskill, NJ: Hampton Press.

Paré, A., & Szewello, H. A. (1995). Social work writing: Learning by doing. In G. Rogers (Ed.), *Social work education: Views and visions* (pp. 164–73). Dubuque, IA: Kendall/Hunt.

Park, P., Brydon-Miller, M., Hall, B., & Jackson, T. (Eds.). (1993). *Voices of change: Participatory research in the United States and Canada.* Toronto: Ontario Institute for Studies in Education.

Polanyi, M. (1964). *Personal knowledge: Towards a post-critical philosophy.* New York: Harper and Row.

Polanyi, M. (1967). *The tacit dimension.* New York: Anchor.

Reid, I. (Ed.). (1987). *The place of genre in learning: Current debates.* Geelong, Australia: Deakin University Press.

Rogoff, B. (1991). Social interaction as apprenticeship in thinking: Guided participation in spatial planning. In L. B. Resnick, J. M. Levine, & S. D. Teasley (Eds.), *Perspectives on socially shared cognition* (pp. 349–364). Washington, DC: American Psychological Association.

Russell, D. R. (1995). Activity theory and its implications for writing instruction. In J. Petraglia (Ed.), *Reconceiving writing: Rethinking writing instruction* (pp. 51–77). Mahwah, NJ: Lawrence Erlbaum Associates.

Russell, D. R. (1997). Rethinking genre in school and society: An activity theory analysis. *Written Communication, 14*(4), 504–554.

Schryer, C. F. (1994). The lab vs. the clinic: Sites of competing genres. In A. Freedman & P. Medway (Eds.), *Genre and the new rhetoric* (pp. 105–124). London: Taylor and Francis.

Segal, J., Paré, A., Brent, D., & Vipond, D. (1998). The researcher as missionary: Problems with rhetoric and reform in the disciplines. *College Composition and Communication, 50*(1), 71–90.

Selzer, J. (1983). The composing processes of an engineer. *College Composition and Communication, 34,* 178–187.

Spilka, R. (Ed.). (1993). *Writing in the workplace: New research perspectives.* Carbondale, IL: Southern Illinois University Press.

Street, B. V. (1984). *Literacy in theory and practice.* Cambridge: Cambridge University Press.

Street, B. V. (1995). *Social literacies: Critical approaches to literacy in development, ethnography and education.* London: Longman.

Tandon, R. (1988). Social transformation and participatory research. *Convergence, 21*(2/3), 5–15.

Vygotsky, L. S. (1978). *Mind in society: The development of higher psychological processes.* Cambridge, MA: Harvard University Press.

Wertsch, J. V. (1991). *Voices of the mind: A sociocultural approach to mediated action.* Cambridge, MA: Harvard University Press.

APPENDIX A:
HANDY PHRASES FOR SOCIAL WORK RECORDING

1. **For establishing authority:**
   - After working with this client for # years, I have come to see . . .
   - After careful consideration, I have reached the following conclusion: . . .
   - Following many hours of discussion with the client and family, I understand that . . .
   - Within Inuit communities there is a belief that . . .
   - The # years that I have worked in social services have taught me that . . .
   - Inuit families value . . .

2. **For acknowledging others and establishing mutual respect:**
   - I know that you are worried/upset/angry/unhappy about . . .
   - I share your concern about . . .
   - I believe that together we can solve this problem if we . . .
   - Like you, I am determined to find a solution . . .
   - If I can help with this situation . . .
   - I would like your help with . . .

3. **For stating uncertainty or possibility:**
   - Perhaps the best thing to do would be . . .
   - It might be possible to . . .
   - Although I am not certain, I do think we might . . .
   - I wonder what would happen if we . . .
   - Could it be that the client is . . . ?

4. **For making recommendations:**
   - May I suggest that we . . .
   - In my opinion, we should . . .
   - My advice would be to . . .
   - In my judgment, we should . . .
   - I recommend that we . . .

5. **To signal a relationship between ideas or events:**

   - As a result of these events, the client has . . .
   - Consequently, there have been problems with . . .
   - First, the family must. . . . Second, they will need. . . . Finally, they could . . .
   - After the child is back in the family, then the parents . . .
   - If the situation continues, then we will have to . . .

6. **To signal concern or danger:**

   - These actions are a warning to us . . .
   - We must use caution handling this case because . . .
   - There will be serious consequences if we do not . . .
   - If we do not act immediately . . .
   - There is a danger that . . .
   - We need to be careful that . . .
   - I am worried that . . .

APPENDIX B: EXCERPT FROM
CULTURE AND COMMUNICATION REPORT

Overall Problem

Both Inuit and non-Inuit workers are expected to begin working as Northern social workers without proper orientation, since there is little training offered between the time they are hired and the time they begin to take on responsibilities. The professional and cultural difficulties they face are great and their supervisors and the community have high expectations. In addition, there is not enough opportunity for on-going training or professional development, and insufficient support from other resources in the community. As a result, workers are not able to perform their professional responsibilities as well as they might and many problems arise. Finally, the hard work and emotional pain the workers must endure often goes unacknowledged and apparently unappreciated, which leads to low morale and burn-out. These three aspects of the overall problem are described separately below.

Orientation and Initial Training: Problems

For non-Inuit workers, the local customs and culture are new and unfamiliar. Their sense of being outsiders or strangers is strong, and they often feel less useful or helpful than they would like to be. Although the non-Inuit workers often have formal training, they have little or no experience in the North and cannot speak Inuktitut. They must often speak to clients through a translator. In the end, they feel a distance between themselves and their clients. Just as important, they feel a distance from their Inuit colleagues, and that makes working together difficult.

Inuit workers have extensive experience with the community, but lack professional training. As a result, they are very familiar with the problems their clients face but unprepared to handle those problems in the ways expected by supervisors. Consequently, they often feel inadequate, especially with supervisors and non-Inuit workers, and many resign their jobs. Although some initial information is supplied in written texts, it is often difficult to relate that material to real life in the community.

Both Inuit and non-Inuit workers must act in accordance with the laws concerning Youth Protection and the Young Offenders Act, but are often unfamiliar with those laws and their local implications. This can lead to dangerous situations.

Orientation and Initial Training: Recommendations

If Inuit and non-Inuit workers are given more extensive orientation and initial training, they will be more effective both immediately and in the long run, as they build on the skills acquired at the beginning of their work. That initial preparation could include:

- More information provided during the hiring process, so that new Inuit and non-Inuit workers would have a better sense of the demands and expectations they will face in the job.
- Clear explanation of all forms, procedures, and regulations that affect day-to-day work in the social services.
- Clear explanation of the laws governing social work practice, especially those concerned with the Young Offenders Act and Youth Protection.
- Introduction to all the services and contact people available in the communities, including CLSCs, women's shelters, foster parent services, and so on.
- Formation of a "buddy" or mentor system, so that all new workers would have support from colleagues.
- Gradual increase in workload under the supervision of an experienced worker, so that new workers would have an apprenticeship period.
- Assistance in forming partnerships with colleagues and allied professionals (doctors, nurses, lawyers, psychologists, etc.), so that team work is possible.

APPENDIX C:
EXCERPT FROM MINISTRY REPORT

Objectivity

Perhaps the most critical recording problem facing social workers is the profession's persistent and pervasive myth of "objectivity," the word most often used to describe the ideal social work record. Although the lengthy and detailed reports of the past were sometimes too biased and often revealed as much about the worker as the client, the contemporary concern with opinion-free, factual, and scientific recording denies the actual situation that workers face: whenever they write about someone's life they must interpret, theorize, speculate, and judge. In fact, most records require social workers to assess, recommend, council, argue, justify, and otherwise present perspectives based on observation, judgement, and personal belief. Truly objective statements are measurable or independently verifiable, and that restricts workers to dates, addresses, telephone numbers, weight, height, and other quantifiable data. Even hair colour and gender may be open to interpretation.

As a result of the impossible demand to be "objective," workers are forced to be unhelpfully brief—recording only those facts that seem unquestionably true but that offer little opportunity for insight to writer or readers—or so vague as to be meaningless. At times, workers employ highly technical language, or jargon, in an effort to sound dispassionate or to conceal their own feelings and impressions.

An even more damaging response to the requirement of "objectivity" is the use of those impersonal forms of writing that are often, but mistakenly, viewed as "professional": "It is believed that . . ."; "The worker recommends . . ."; "Assessment of the situation indicates that . . ." Such language masquerades as objective and detached but, in fact, simply disguises interpretation and raises opinion to the status of certainty.

This is no trivial matter of style, but an issue that goes to the heart of the profession's authority. How should workers speak and write about what they know? On what basis are social work observations made? What claims support a social work perspective? Should the profession aspire to the antiseptic, impersonal language of science, or acknowledge that social work involves interpretation, hunch, intuition, and speculation?

Solution

This is a fundamental issue, and one not easily solved by individual workers or even whole agencies or other institutions. However, like most of the problems described in this report, it can be alleviated by an open and ongoing discussion of the place, purpose, and problems of recording. This is a refrain throughout the recommended solutions that follow: people involved in social work must put recording on their organizations' agendas. They must talk about it, share ideas and expertise, develop collective standards and expectations, provide workers with the support and resources they need, and otherwise treat recording with the serious consideration it deserves as a central and difficult part of social work practice.

Naturally, workers must strive for unbiased records, but the effort to achieve "objectivity" in any kind of scientific sense—that is, without personal judgment, values, or beliefs—is doomed to failure in social work writing, where workers *must* interpret actions, make assessments, offer recommendations, speculate about causes, and otherwise exercise judgement. Even apparently "objective" descriptions of clients or the inclusion of verbatim quotes involves choice and, therefore, judgement. From *everything* that could be said about someone, from *all* that a client has spoken, the worker must select.

The first step toward a reduction of this problem is to face it and, through discussion, to reach a consensus or collective understanding about what constitutes fair and professional recording. Within agencies, hospitals, and other institutions where social work operates, workers must articulate their expectations about how personal bias, opinions, values, attitudes, and other personal or "subjective" perspectives affect recording.

Active-Practice: Creating Productive Tension Between Academia and Industry

STEPHEN A. BERNHARDT
University of Delaware

It may be tempting to wish for close relations of academia and industry, but we should be careful what we wish for. We should not immediately assume that the best case would be to have full concord of goals and work practices between academia and industry. To some extent, the two worlds are best left to their own purposes and tasks. Many academics would challenge the notion that schools should be more accommodating to workplaces, fearing that the curriculum might become dominated by the goals and practices of the workplace. Many academics also welcome some degree of distance for the critical stance it affords toward workplace practices that an informed (and socially attuned) rhetoric delineates.[1]

Dicks (chapter 1, this volume) examines reasons for maintaining this distance. He demonstrates how both industry and academic professionals have many good reasons to keep their counterparts at arm's length. A desirable goal in the field of technical communication might be to achieve a productive tension between academia and industry. In this chapter, I make a case for why academia and industry remain separate in their goals and practices, and then argue that their relationship could improve significantly with the shaping of what I call

---

[1] The technical and professional communication literature well recognizes the need for critical distance from industry (see, for example, Herndl, 1993; Spilka, 1993; and Thralls & Blyler, 1993).

active-practice, an approach that involves educators and practitioners working together through project-based activities to achieve more fruitful and fulfilling working partnerships.

Academics and industry practitioners have a long way to go in their ability to work profitably together toward shared perspectives, goals, and work practices. Academia has not yet built strong enough ties with industry, partly because of limitations imposed by the inherited culture of English departments and partly because of the structural conditions determining the lives of academics within writing programs. Likewise, based on certain cultural assumptions, industry has tended to operate with an incomplete perspective on technical communication programs in academia and their potential relevance to commercial settings. This section examines cultural factors that have led to distinct academic and industry perspectives that discourage a close academic–industry alignment.

THE ACADEMIC PERSPECTIVE

For a variety of historical reasons, academic practice has become segregated from workplace practice. A major reason for the continuing separation is the alignment of technical communication programs within traditional English departments. A second set of reasons involves the social and material conditions of working inside a university.

The strong commitment to the humanities and to tenets of liberal education within universities, and within English departments in particular, helps maintain practices at odds with the workplace. Many whose traditional academic home is within the liberal arts resent the increasingly careerist and market-oriented trends in education of the past 20 years. Garay (1998) is particularly compelling in her analysis of why some English teachers tend to position themselves in opposition to workplaces, citing what she terms a *filthy lucre complex* and a general resistance to allowing education to become more careerist/vocational and less humanistic/educational. She is also quite resourceful at suggesting ways to reconcile unproductive tensions and serve our students' best interests.

Academics tend to fix inwardly on their classrooms, programs, and texts. Teaching has long been recognized as an isolating profession, conducted with younger people behind the closed classroom door (Lortie, 1977). Technical communication, then, finds itself embedded within a larger culture of English and Humanities studies that either actively or passively resists thinking about work. Because of a long tradition of separate and isolated practice, schools can ignore work and still do their business.

Assumptions about identity formation underscore the divide. English departments, in particular, maintain a commitment to narrative, essayistic, and personal texts, under the assumption that identities are construed within the realm of personal experiences and meanings. Working adults, in contrast, use the language and communicative practices of their work lives to establish who they are (Wenger, 1998). For most adults, the workplace is the primary site of identity formation and literacy development: Most of the texts that are read and written on a daily basis are done so in work contexts. When adults say who they are and talk about what they do, it is the work context that predominates over home or community.

There is a broad perception among many academics that separation of industry and the academy is a good idea. Many whose traditional academic home is within the liberal arts resent the increasingly careerist and market-oriented trends in education of the past 20 years. A common academic argument is that a public education, which for the most part is what universities offer, should not be dictated by private industry needs and that some separation is vital between education and vocational training. We should not underestimate the strength and importance of this resistance.

Academics in technical communication are generally more in tune and in touch with the workplace. Yet, as MacNealy (1999) found in her survey work, many technical communication programs actually live in tense opposition to their home departments. Furthermore, there are issues associated with daily life in university departments that conspire to keep the focus of many academics internal:

• Academics tend to be busy in their academic lives with creating and running programs, training new teachers, running professional organizations, creating support for writing within their institutions, providing support for technology, and working closely with students in time-intensive tasks.

• Teaching loads tend to be higher in English departments and technical communication programs than in sciences or engineering. The teaching tends to be time-consuming, including grading papers, having one-on-one conferences, and working with high loads of independent study, student theses, and dissertations.

• Departments in liberal arts are underfunded in terms of salaries, equipment, and lab budgets. Whereas other departments have research assistants, faculty in English or technical communication tend to have teaching assistants, who teach their own classes, not truly assisting. Academics in professional communication lack the research time, support, and infrastructure that would support extended engagement with industry.

• The grants that are most available in technical and professional communication are centered on teaching, assessment, or training and development: for example, these grants might encourage minority students to enter careers where they are underrepresented, or might support Writing Across the Curriculum (WAC), workforce training, and program development. External funding for research on industry concerns is difficult to find.

Essentially, technical communication faculty are structured into a teaching environment that isolates them from industry and its concerns. Academic environments are structured to provide a teaching faculty that focuses on developing students' communication skills, with only modest expectations for research in the time left after teaching and service, and with little affordance and only rare rewards for project-based interactions in other workplaces. There are few incentives to make connections to workplaces.

THE INDUSTRY PERSPECTIVE

Industry practices similarly keep practitioners separate from the academy. English departments and their programs in technical and professional communication are typically not even on the map for most industry representatives. Each term on many campuses, industries schedule visits during recruitment fairs for interns and hires, but they look for specialized majors such as electrical engineers and computer programmers, who represent well-defined areas of need. Technical communication enjoys this status only within a highly restricted set of companies, typically computer companies.

Conflicting industry and academic perspectives on training are well exemplified in charged discussions of what skills graduates need. Too often, needs are defined in terms of specific software skills—technical communication graduates should know RoboHelp or FrameMaker, for example. Hailey (1997a, 1997b) identifies job demands based on his analysis of such sources as job ads, but there are problems with arguing from job ads about what students need to learn the most.

Interestingly, the engineering field is not asking for highly specific training based on anticipated job demands, but for a broad education including strong communication and collaborative skills, analytical ability, a systems perspective, and an appreciation of cultural diversity within a multidisciplinary framework (American Society for Engineering Education, 1994). Another report from the same organization (Panitz, 1998) suggests that an engineering education is fast becoming "the new liberal arts degree." It would be ironic for technical and professional communication to pursue increasingly specialized practice centered on narrow technical skills while engineering pursues a broad curriculum and skill set for its students. An insistence on the part of industry on closely defined, "necessary" skills for technical communicators works against the broad role of the university to give students deep understandings in rhetoric, communication, and information design.

Overall, industry could do more to help students develop a vision of possible work roles and build identities as working professionals. Far too often, high school and college students are employed primarily as cheap labor in unskilled service roles. These students work within primarily unreflective modes, discovering in work a relentless routine of the ordinary. A more thoughtful society would put young people in positions where they gain important skills, develop critical understandings of work, and come to understand meaningful career choices. This happens in the European model, in which many students move quickly from gymnasium at age 19 or 20 into apprentice work situations, where they continue to develop a broad range of highly sophisticated work skills. Industry could do more to offer richer apprenticeship opportunities for students (National Center on Education and the Economy, 1990).

Industry also tends to split from the academy in terms of the place or value of technical communication research. Practitioners have little time to aggressively research their own conduct. Even in a research-intensive setting, such as in pharmaceuticals, they take an unreflective, experience-based approach to making decisions about best practices regarding document development. Rarely does a company investigate in structured ways such issues as how much documentation costs, how effective it is, how it compares across industry, and how it might be tested and improved.

Privacy norms in industry also limit the potential for research partnerships between the academy and industry. There are limits on how specific external reporting may be, and industry takes many issues off the table for discussion outside a company.

SHAPING AN ACTIVE-PRACTICE BETWEEN ACADEMIA AND INDUSTRY

Over the past few decades, despite many opportunities to establish closer working ties between academia and industry, progress toward this goal has been disappointing. Here I argue that a useful and powerful way to improve relations between academia and industry would be to establish shared communities of practice involving frequent, active, project-based cooperation. The goal of an active-practice is to engage practitioners and academics in a range of projects. By working together on shared problems, practitioners and academics can reshape practice in both spheres.

The theory behind active-practice is that people do not understand and appreciate each other until they work together toward shared goals. Academics and industry practitioners are unlikely to appreciate each other until they actively develop a shared practice through working together. What is needed is a community of practice (Wenger, 1998) that bridges the workplace and academia—a shared sphere of activity (therefore an *active-practice*) that would allow a theoretically and empirically informed practice to emerge. Without such an active-practice, we are likely to continue to engage in unproductive behaviors, with industry panels telling schools what they should be doing or university academics writing abstract attacks on business.

To explore what an active-practice might look like, we can identify the following examples of productive relationships between the computer industry and academia:

- Well-established intern programs.
- Specialized coursework in computer documentation.
- Campus visits from high tech recruiters, who know the technical communication programs and seek their graduates.
- Substantial and highly productive applied research on the design of manuals and tutorials, on the transition of documentation from paper to online, and on users and usability.
- Highly developed research laboratories to support the study of computer documentation.
- Professional societies (such as the Society for Technical Communication) bringing together participants from both sides for special meetings, annual conferences, and the like.

As a result of this type of collaboration, the specialty of computer documentation has established a well-defined research literature with highly developed constructs built on active work practice (e.g., minimalism, task orientation, usability testing). The specialty is now well supported by a wealth of publications, with a lively critical/theoretical debate as to goals, practice, and methods (e.g., Johnson, 1998; Johnson-Eilola, 1997). Students or faculty interested in moving into the computer industry have options for interning and can meet campus recruiters. Here is a place where academia and industry work actively together, joining their separate spheres of practice into a single, shared active-practice. The same model of active-practice could be enacted in other industries. In an active-practice, individuals are able to move back and forth between being academic professors and working in industry settings; likewise, industry practitioners are able to spend time on campus, as through IBM's program that loans employees to universities, where they teach and work with students.

In a successful active-practice, industry would identify and propose research projects for collaborative investigation. Certain academic programs and companies would establish close relations with some flow of resources. In the best instances, academia and partner companies would collaborate toward shared goals, actively sharing information, working on projects, and enjoying close intern and recruiting relationships. Students could construct work identities and faculty could gain first-hand experiences that allow them to imagine life in a corporate setting. In turn, corporate representatives would come to understand the values and constraints of university programs. Both parties would see their interests as intertwined.

WHAT ACADEMICS COULD DO TO FACILITATE ACTIVE-PRACTICE

Academics will likely need to lead the way, developing and asserting their value to industry. The potential sites for engagement are everywhere. There is much that faculty might do to make the initial move:

- Teachers might bring in the world of texts that exist in commercial spheres.
- Teachers might make certain that students are moving to and from academic and work settings to experience, compare, and develop wide-ranging opinions.
- Students could be expected, in both secondary and postsecondary classrooms, to investigate various work settings, learn a language of business and technology, and understand the logic of products, materials, logistics, production, finance, marketing, sales, and information systems.
- School projects might be approved based on a cost–benefit analysis, and they could run on schedules and within budgets.
- Industry representatives could mentor students or evaluate projects.

Thinking in such terms and working with such materials could become the norm, rather than the exception.

An activity-based practice could move across boundaries and build bridges through specific practices. An activity-based practice would need to include the following:

- A robust theory that attends to cost–benefit analysis and resource issues.
- Research partnerships that address industry-specific problems (those with real consequences for industry).
- Cooperative work through professional societies.
- Training and consulting work.
- Increased reflective experience of students in work setting.

Academics could also suggest areas for active-practice projects. What academics know and do could have immense value for industry, assuming it can be presented and practiced in ways that make apparent its usefulness to industry. Academics know much about document prototypes (models), usability testing (writing review and revision), quality standards (writing assessment or grading), document project management (getting work in by deadline), and knowledge management (writing as a learning tool). Academics could provide training in process approaches to writing and in strategies for transferring those processes to work. They are also well positioned to investigate transfer of learning, outcomes of training, roles of writers in organizations, and achieving process change in work settings with regard to communication practices.

Active-practice could similarly be a means for academics to work with literacy concerns among frontline workers. Current initiatives, funded by the federal and state governments, unions, and industry, are directing unprecedented resources into welfare-to-work programs, in-plant technical training, workforce literacy, and other areas. The need for improved technological literacies in various industries has never been stronger. Working with frontline workers to give them "new workplace" skills could help the academic profession enact a politically informed vision, through service, to those who most need and deserve training. Academics would find their talents and perspectives welcomed by those doing the difficult work of shaping meaningful programs of education and training for a rapidly changing workplace.

Academics have also developed impressive expertise in writing review and evaluation, but have yet to port this expertise to industry settings. The literature on writing assessment is directed almost exclusively at testing student writing: the writing produced in classrooms and shaped by academic norms. Academic researchers could also work profitably with industry practitioners to develop quality standards for documentation, usability testing, and measures for improvement in document quality that allow industry to assess progress.

Cooperative practice between industry and academia would revitalize university research programs. Because considerable time must be devoted to developing partnerships, academics would need to figure out how to make the time, buy off the course releases, and hire graduate student researchers. Perhaps STC could revive earlier efforts to work with academic and industry researchers to develop a clearinghouse for cooperative research partnerships.

WHAT INDUSTRY COULD DO TO FACILITATE ACTIVE-PRACTICE

Industry could take the lead in sponsoring research programs that would add value to specific industries. STC, for example, sponsored a project that asked about the value that publications groups bring to a development environment (Ramey, 1995; Redish, 1995). Other fertile areas of research are described elsewhere in this volume. Sponsoring academic research is a familiar activity in the sciences and engineering, and industry is well aware of the advantages that accrue when research is conducted through an outside laboratory. Although industry sponsors scientific, medical, and technical studies, it has yet to make a similar connection to technical communication.

For research sponsorship to work well, industry needs to find ways to employ the research talents of academics without shutting down the academic ethos of shared knowledge. They need to work harder to separate knowledge that is truly proprietary and actually promises a competitive advantage if kept private, from other kinds of knowledge that can be shared and talked about without compromising an industry's advantage. The development of minimalist practice, sponsored largely by IBM and involving academic alliances, was based initially on research done within an IBM lab (Carroll, 1998). It proved to be work of such high interest as to attract attention from various quarters in both industry and academia; cosponsored meetings were held for exchange of papers and perspective; and edited volumes were coauthored across the industry–academic line. The work developed continues to benefit industry as a whole, and not just IBM in particular. It also gives academic programs a fine example of practice-based research, where the questions asked can have a large impact on the design of texts.

THE VALUE OF PRODUCTIVE TENSION
BETWEEN ACADEMIA AND INDUSTRY

Academia and industry need to collaborate on cultivating a shared vision of active-practice. It is primarily by working together on projects of shared interest that they will build stronger ties. Every attempt at communication across boundaries, such as working together on research and education or planning and executing events together, has the potential to create deeper understanding and cooperative work relationships. We should see to it that the boundaries of industry and academia are crossed frequently and with an open mind as to what might occur. The closer relationship we desire will follow.

There are incentives on both sides for pursuing an active-practice. For academics, industry-based collaborative projects bring credibility to professors and programs, with valuable sources of experiential learning for students. On the industry side, an active-practice would bring to technical and professional communication the same benefits enjoyed in other areas of industry–academic collaboration. Access to scholars means access to research-based knowledge and methods, which can be applied to meaningful industry-identified concerns. Alignment with programs provides a stream of students as interns and hires, and ensures that industry practices are based on current models of effective documentation and communication practices.

Working together, merely as partners, without presumptuous assumptions on either side, would bridge divides of values and perspectives. Without telling academics what to do or how to run their programs, industry would have influence. In turn, academic perspectives on industry—whether theoretical, empirical, or applied—would change in positive ways from the benefits of working together with engaged workplace personnel in industry settings on actual problems. The result would be a tempering of distant, academic critical posturing and industry skepticism, together with a recovery of relevance and understanding across the divide. The relationship would be kept in tune by the natural tension of working together.

The rewards will be extremely worthwhile: The work that is accomplished will invigorate the field of technical communication.

REFERENCES

American Society for Engineering Education. (1994, October). *The Green Report: Engineering Education for a Changing World*. Available: *http://www.asee.org/publications/reports/green.cfm*

Carroll, J. M. (Ed). (1998). *Minimalism beyond the Nurnberg funnel*. Cambridge, MA: MIT Press.

Garay, M. S. (1998). Of work and English. In M. S. Garay & S. A. Bernhardt (Eds.), *Expanding literacies: English teaching and the new workplace* (pp. 3–20). Albany, NY: SUNY Press.

Hailey, D. (1997a). Examining industry need and building a multimedia program to match. *Text Technology, 7*(4), 1–18.

Hailey, D. (1997b). What do you need to be needed? *INTERCOM*, 26–29.

Herndl, C. (1993). Teaching discourse and reproducing culture. *College Composition and Communication, 44*(3), 349–363.

Johnson, R. R. (1998). *User-centered technology: A rhetorical theory for computers and other mundane artifacts*. Albany: SUNY Press.

Johnson-Eilola, J. (1997). *Nostalgic angels: Rearticulating hypertext writing*. Norwood, NJ: Ablex.

Lortie, D. C. (1977). *Schoolteacher: A sociological study*. Chicago: University of Chicago Press.

MacNealy, M. S. (1999). Can this marriage be saved: Is an English department a good home for technical communication? *Journal of Technical Writing and Communication, 29*(1), 41–64.

National Center on Education and the Economy. (1990, June). *America's choice: High skills or low wages*. The Report of the Commission on the Skills of the American Workforce. (National Center on Education and the Economy, 39 State Street, Suite 500, Rochester, NY 14614. Tel. 716-546-7620).

Panitz, B. (1998, November). Opening new doors. *Prism online*. Available: http://www.asee.org/prism/november/html/opening_new_doors.htm

Ramey, J. (1995). What technical communicators think about measuring value added: Report on a questionnaire. *Technical Communication, 42*, 40–51.

Redish, J. C. (1995). Adding value as a professional technical communicator. *Technical Communication, 42*, 26–39.

Spilka, R. (1993). Influencing workplace practice: A challenge for professional writing specialists in academia. In R. Spilka (Ed.), *Writing in the workplace: New research perspectives* (pp. 207–219). Carbondale: Southern Illinois University Press.

Thralls, C., & Blyler, N. R. (1993). The social perspective and professional communication: Diversity and directions in research. In N. R. Blyler & C. Thralls (Eds.), *Professional communication: The social perspective* (pp. 3–48). Newbury Park, CA: Sage.

Wenger, E. (1998). *Communities of practice: Learning, meaning, and identity*. Cambridge, UK: Cambridge University Press.

II

RE-ENVISIONING THE PROFESSION

New directions have characterized the steady evolution of technical communication from its earliest days. Yet, at the risk of echoing prior claims, we believe that the new directions proposed in this volume are categorically different and especially exciting.

In other periods, new directions and progress centered on some aspect of the communications that we create. It was a breakthrough, for example, when we transformed the activity of "translating" subject matter into the goals of writing and designing for audiences, contexts, media, and purposes. The outcome was human-centered documents, task-oriented manuals, and the like. Another leap forward, still underway, involves mastering effective communications in new media, and moving beyond print to hypermedia, video, animation, virtual realities, and numerous combinations.

In this volume, the call for new directions diverges from this primary concern with communication products themselves. Instead, as Part II chapters insist, we need to broaden our vision, goals, and concerns to new categories, beyond the communication products that we develop. We need to take on boundary spanning roles organizationally and become the originators and purveyors of information and knowledge crucial for strategic planning and decisions about product directions. Anscheutz and Rosenbaum (chapter 9) aptly position the profession at a crossroads. We can choose to define technical communication traditionally, focusing on knowledge and skills for putting out our products, or we can choose a more expansive definition and professional identity, one that embraces a network of activities within and across our organizations. This network of efforts involves strategically and tactically affecting the usability, usefulness, and quality of technologies. To do so, the development of our portion of a technological product is only one activity among many others. We also must contribute to product planning, design, corporate decision making, process innovation, and the rethinking of old assumptions.

To move ahead, to grow as a profession, and to have greater impact on the uses and understanding of the technologies about which we communicate, the authors in Part II agree that we need to redefine our field and identities more expansively. We need to shift from defining our work and value primarily in terms of the communications we produce to focusing just as strongly on the strategic value to what we add to our work contexts and product designs and directions. Only through this larger role can we significantly affect technological usefulness, usability, learnability, enjoyment, and quality.

In making this shift, we need to embrace many roles. In our technology, knowledge and information are highly dynamic. We need to take on and adapt to fluid roles as we create, disseminate, and use information to influence the directions and uses of technologies. As with most reforms, the articulation of this group of authors reflects tacit trends that have been occurring for several years, for at least a decade. The Part II chapters seek to give voice to these trends toward broader and more influential roles and to formally investigate how the profession may build on strengths and opportunities for a more expansive future. When combined, this group of chapters makes a clear case that the time has come to embrace this transformation formally as a core part of our professional identity.

One main issue that arises from this group of chapters has equal significance for practitioners and academics alike. It is the need to strike the right level of detail in redefining our field and identities more broadly. The level must not be so abstract and broad that it is all things to all people, yet not be so narrow and concrete that it is completely context-specific or tied only to how-to practices. We need to strike this balance to assure that we neither overextend our reach beyond our strengths nor underdetermine our unique value across work contexts in the university and industry.

Technical communication is not the only field vying for a seat at the strategic table. In the past decade, specialists trained in university programs in human–computer interaction, information science, knowledge management, industrial engineering, cognitive psychology, and corporate anthropology have all taken on the expansive roles that technical communicators are moving into. These include roles as usability analysts, user experience researchers, content strategists, and interaction designers. As Spilka in chapter 6 suggests, because of this competition, defining and acting on strengths and unique contributions that extend beyond our immediate communication products may be crucial to survival. Therefore, it is important to get the level of detail right in articulating our vision, goals, direction, roles, and functions.

This group of chapters does not explicitly address the issue of level of detail, but implicitly the chapters evoke it and raise some challenging questions about how to define visions, goals, and the like. They offer questions and insights that should spark ongoing dialogues and achievements in the field for many years to come, such as these:

To what standards of practice, quality, and measurement can we commit ourselves? These standards need to transcend the boundaries of our workplaces, yet remain within our unique scope and expertise. For standards of practice, quality, and evaluation measures, Part II authors suggest looking into the following areas that are relatively neglected at present and need attention:

- Creating a long-term investigative agenda centered on the most critical problems associated with improving and enhancing humans' interactions with and communications about technologies, problems that involve real world initiatives and are pressing and meaningful to academics and practitioners alike.
- Disseminating knowledge, insights, and contributions within the field and outside of it in ways that address targeted readers' concerns, styles of reading, and expectations.
- Defining the scope and content that we can contribute to strategic knowledge, processes, and decisions in our organizations and in the organizations of our readers or users.
- Characterizing and evaluating the leadership that we demonstrate as we vie for positions of organizational influence.
- Developing and providing the mentoring, education, and models of education that foster expansive careers.

For the profession generally and for specific aspects of it, what visions need to inspire the field and be turned into realities? The Part II authors cover a range of visions related to:

- Assuming roles as hybrid professionals that are fostered in education and work.
- Assuring improved usability by assuming a role in every phase of the development cycle.
- Effectively interpreting and disseminating what we do to the diverse audiences who are instrumental to according us value.
- Creating adequate and appropriate organizations for embodying new directions for expansive roles and definitions.
- Narrating unifying stories through real-life cases and through myths to generate a culture of technical communication that transcends current academic and industry boundaries.

In moving toward a more expansive definition and identity, the issue arises whether the term *technical communication* suffices for our profession. Authors variously ask, for example:

- What do we call ourselves? Is this a survival issue? If many fields are competing for the same value and influence, does naming take on a greater significance for us?
- What career paths do we encourage and how do we assure that people continue to identify with and support technical communication professionally?
- In merging product-making abilities with abilities to imagine, design, innovate, and reengineer processes, are other, more hybrid names more appropriate?

Underlying all these questions is the overall issue of value, and it unifies the Part II chapters. Expanding the definitions of our professional work and goals necessarily involves building an image of technical communicators as valuable to the business conducted in our workplaces—as more than just makers of communication deliverables. This perceived value is a complicated dynamic. It has to come about internally through technical communicators' own awareness, aspirations, and experiences of efficacy and credibility. It also has to come about externally, by winning the respect of our teammates, colleagues, managers, audiences, and the public at large. The Part II authors explore this issue of value and new direction from many angles.

Rachel Spilka's chapter looks internally toward affirming our value. She believes that we have to treat our diverse roles and functions as strengths, not weaknesses, and strive to evolve a full-fledged profession. As a profession, we will offer shared visions and goals and an expanded definition of technical communication. Professionalization, Spilka argues, is a desirable route toward value, status, and prestige. As a vehicle for building and enacting these shared visions and goals and their corresponding value, Spilka proposes creating a new organization, a consortium designed specifically for moving the field in new directions. She envisions this organization including academic and practitioner voices equally.

Building on the idea that currently we do not realize our full potential as a field and a profession, Karen Schriver (chapter 7) suggests one pressing reason why. We do not spend enough time and effort revealing the value of our studies and investigations to the wide range of stakeholders for information design research. Schriver summarizes the rich and varied investigations that have been conducted in information design over the years. However, by and large, dissemination of these investigations has been insular. As Schriver argues, research rarely goes out in forms that make an impression or a difference to crucial internal stakeholders such as practitioners and teachers of technical communication and related fields, let alone external stakeholders such as consumers, the mass media, or other public citizens. This lack of dissemination is stunting the growth and perceived value of the field. She urges better, increased communication across constituencies. Writing effectively to varied constituencies, however, is rhetorically challenging. Schriver describes specific traits about each of these stakeholder audiences to guide investigators in communicating with them. In appendices, she also provides, as models, "code-switching" examples of actual communications about the same subject framed differently to appeal to distinct audiences.

Brenton Faber and Johndan Eilola-Johnson, like Spilka, believe that technical communication lacks professional qualities, which is hurting its identity and value in the workplace. In chapter 8, they examine the profession against the dynamics of our global economy and argue that if we continue to define ourselves by the products we create, we will be calamitously out of sync with larger economic trends. These trends include customer or market-driven production processes and inventive uses of knowledge and design to differentiate goods and services in competitive markets. To fit within these trends, Faber and Eilola-Johnson urge us to envision ourselves as "hybrid professionals" who combine product knowledge and strategic design and business knowledge. To move in this direction, they propose new models of education implemented in hybrid institutions. These institutions differ from more common industry–academic endeavors such as corporate universities or collaboratories. They are distinctive in the novel ways in which they distribute and integrate learning, knowledge, and the flow of information across university and industry boundaries. Faber and Eilola-Johnson describe their vision of these hybrid institutions, their model of education, and their potential for moving the profession forward by drawing on examples from their own experiences.

Lori Anscheutz and Stephanie Rosenbaum give concrete form to the theme of more expansive career paths by narrating the experiences of six professionals who have succeeded in achieving the strategic value in their companies that Part II contributors urge for the profession. Anscheutz and Rosenbaum trace these professionals' moves from early roles in traditional technical communication areas such as documentation, to positions with more influence over the design and direction of the whole product, for example, associate partner for technology in e-commerce solutions, usability labs manager, and business operations strategist. The authors are concerned that these career evolutions signal a move out of technical communication, and they want to avoid any severing of ties. To do so, the profession needs to establish expansive visions and goals so that people remain committed to affiliating with the communication discipline long after they lose such terms from their job titles as "information developer," "technical communicator," "writer," or "editor." Anscheutz and Rosenbaum cite current patterns that suggest a continued and increasing trend toward career transitions. With this future in mind, they couple their case histories with suggestions for what specialists in both industry and academia can do to mentor and support evolving careers.

Barbara Mirel, equally concerned that technical communicators move into roles of influence in software design and production, proposes in chapter 10 a vision of usability for professionals to advance in order to assume leadership roles in building usability into software from its inception on. Although many technical communicators currently are making a transition into usability roles, usability leadership is still sorely lacking in the computing industry. It is especially lacking when it comes to moving beyond ease of use issues and building products that are truly useful for the complex problems and tasks that comprise a good deal of users' everyday work. Mirel explains why conventional ways of analyzing users' needs and modeling their tasks in context will not move software teams forward in building for this usefulness. Nor will current methodologies of development achieve that goal that exclude usefulness from front-end decisions about product scope, architecture, and features.

To assume leadership roles in usability and improve the usefulness of products, Mirel urges technical communicators to advance new approaches to task analysis and development processes. Achieving strategic value as a field is intricately tied to articulating and implementing a vision. Toward these ends, Mirel believes that professionals in industry and academia need to investigate the types of problems that provide visionary insights and bring the value of these insights to bear on their respective worlds.

Concluding Part II, Russell Borland offers a narrative essay comprised of tales and commentaries. Offering a mix of trickster tales, creation myths, and parables, Borland's stories capture the journey of technical communicators over the years—the travails that we have encountered, the epiphanies that have led to new transitions, and the continuing challenges that beckon us toward expanded identities and new measures of value. Underlying the analysis and debates that go on within and across the worlds of academia and industry about our field and its future are narrative tales. These are the stories that we tell ourselves and present to the outside world about who we are, how we work, and what we contribute. They have long-lasting power and persist far after the arguments fade away. Borland weaves the same vision of enhanced strategic value into his narrative essay that we find in other Part II chapters. Here it takes on "deep structure" qualities, serving as a cultural narrative that runs through the profession, binds the academic and industry worlds, and inspires our journey toward the future.

6

Becoming a Profession

RACHEL SPILKA
University of Wisconsin–Milwaukee

The field of technical communication is suffering an identity and credibility crisis. Ask any technical communication specialist, in industry or in academia, to list uncertainties about this field, and almost inevitably, critical issues of identity, credibility, and value emerge, such as these:

- What should we call this field? Why have so many different titles emerged?
- How should we define the field? Why has the field had so much difficulty reaching consensus on a definition?
- What gives rise to the need for technical communication? What do most technical communicators do? What underlying problems in industry need the specialized help of technical communicators?
- Given their potential contributions to organizations, why do technical communicators have so little influence in strategic decisions in their organizations? To what extent is this a crucial source of dissatisfaction for technical communicators?
- What do technical communicators wish they could be doing? Why are so many dissatisfied with what they are doing now?

These uncertainties have characterized our field for quite some time, but could also be regarded as opportunities for significant reform. They could reflect our field's healthy, ongoing struggle to come of age, to evolve into something more permanent, credible, and valued, namely a profession.

In this chapter, I argue that technical communicators need to take the following steps to move the field beyond its current uncertain status and identity so that it can become a profession:

1. Embrace and promote our diversity and recognize that a lack of unity over what to call ourselves and how to define our roles could be where our greatest strengths and contributions lie.

2. Acknowledge that the field is not yet where it needs to be and articulate a vision statement, along with a comprehensive list of goals for the field.

3. Create a new organizational consortium, consisting of members who represent diverse aspects of the field, for the dual purposes of achieving a consensus about the future and coordinating efforts to work toward our vision and goals.

WHY SHOULD WE STRIVE TO BECOME A PROFESSION?

There has been some debate about whether technical communication is a field[1] or a profession[2] (Savage, 1999, describes this in some detail). I define a *field* as a body of knowledge and research and a history of practice that center on a common purpose. Technical communication would qualify as a field because it is based on at least half a century of theory, research, and training and a much longer history of practice at work sites. Perhaps the most common, shared purpose of the field, as reflected in most definitions of technical communication over the past 30 years, is to make workplace information accessible to and useful for a target audience.

I contend that a *profession* shares all these features, but differs from a field in the following ways:

• A profession enjoys universal recognition—a consensus perspective—of its title, definition, features, responsibilities, goals, and standards. For example, throughout industry and academia, just about everyone recognizes what an engineer is trained to do, what an engineer typically does, and the quality of work, or standards, that an engineer is expected to achieve.

• A profession is also characterized by a systematic means of approaching and evaluating tasks, doing the work, determining and then achieving goals and standards, and judging the quality of work against those goals and standards.

• A profession consists of a well-defined community or culture that is characterized by a clear vision and that motivates, inspires, and guides members.

• A profession formulates its most important decisions and policies through organizations, which guide, represent, defend, and support workers within the

---

[1]In academic discussions, a field is often referred to as a discipline. Because the concept of a discipline tends to be used by academics and not by practitioners, I decided not to define it or to compare it to a field or a profession in this chapter.

[2]Some consider a profession to be the same as a "skilled trade," but for the purposes of this chapter, I approach a "skilled trade" as a specialty and contend that both a field and a profession are typically comprised of multiple, diverse skilled trades or specialties.

profession and allow diverse voices in the profession to join, be heard, and matter.

• Members of a profession generally enjoy status, prestige, and power both within particular organizations and among the general public. There is typically little to no debate about the importance and right of workers in a profession to do their jobs as they deem best and to stake a claim in important decision making in an organization.

Technical communication has yet to achieve any of these goals. It is not as mature as professions such as engineering, medicine, and architecture, which have articulated visions for the future, set comprehensive goals and standards for professional work and conduct, and established systems for evaluating the work of practitioners to determine whether those standards have been met. Also, although the largest and most influential organization in the field, the Society for Technical Communication (STC), currently with about 24,000 members, has developed a mission statement along with a limited set of goals and has guided the field in positive directions, it has been unable to help the field evolve into a profession. STC has established committees to look into the possibility of taking steps toward professionalization, including accreditation of academic programs and certification of practitioners, but it has been unsuccessful in achieving a consensus about whether to even strive for those goals (Davis, 2001; Savage, 1999). Also, unlike practitioners in most professions, many technical communicators lack status, prestige, and power at their work sites and are far from reaching their potential. In addition, for the most part, as Schriver argues in this anthology, the work of technical communicators remains mostly unrecognized and misunderstood by the general public.

Davis (2001) argues that the field's lack of consensus about professionalization is a key reason that it has been unable to evolve beyond its current status as a field. Johnson-Eilola (2001) agrees that "our field is just too young and varied to have reached agreement on both academic and practitioner interests, and has yet to develop a standard literature that people learning and practicing in the field all know" or to develop "standard development processes even for relatively common activities such as developing procedural instructions or analyzing audience." Savage (2001) also points to a lack of consensus as a key obstacle to its becoming a profession:

> Technical communication still hasn't achieved full professional status. The argument diverges on the question of whether we will ever become a "true" profession in the traditional sense, and diverges again on the question of whether we should even want such status . . . long before a field can achieve the status of a profession, its practitioners must begin to identify themselves and conduct themselves . . . as "professional" and hold other practitioners to high standards.

More specifically, professionalization would help us move closer to the following important goals:

- Greater work fulfillment for individual technical communicators, who would come closer to reaching their potential in what they can contribute to workplace practice.
- Wider recognition and greater status, prestige, and influence for technical communicators both within particular organizations and in society in general.
- More efficiency and success in defining, measuring, and achieving a high level of quality in technical communication processes and products.
- More chance for the field to strengthen and survive into the 21st century as technical communication begins to compete with other fields for a stake in information development, design, and maintenance.

TOWARD PROFESSIONAL STRENGTH AND PRIDE: SEEING AND EXPLOITING THE POSITIVES IN DIVERSITY

An apparent obstacle to professionalization has been our chronic dissatisfaction with our status as technical communicators. Our own concept of our potential is not mirrored in the roles and influence we have in our organizations and larger social communities. I argue that to elevate our status and to help the field mature into a profession, we need to regard our diversity not as a lack of consensus and unity, nor as a barrier to progress, but rather as a source of strength and pride to promote to others.

Technical communicators are often frustrated by organizational and professional obstacles that seem to stand in the way of meaningful contributions, influence, respect, and professional development. When discussing the status of the field, technical communication specialists often contrast a relatively disappointing current reality with ideal scenarios of the future. For example, note how, in the following quotation, Davis (2001) dramatically contrasts the present (in which she describes technical communicators as "tool jockeys" in a "servant role") and the future (in which she hopes that technical communicators, among other things, will contribute to the "growth of the profession"):

> unless technical communicators want to remain in a servant role, we must become more than tool jockeys. We must complete the evolution from craftsperson to professional. . . . To the extent that we understand and integrate information technology in all its aspects into our programs, we can be successful in contributing to the growth of the profession and to the greater status and capabilities of our graduates. If we remain locked into the image of educating writers who lack the broad-based technical grounding to succeed in the next 50 years, then we are shortchanging the future of technical communication. (pp. 139, 143)

Note the same divergence between the recent portrayal of technical communicators as "translators" (of information and ideas) and an ideal, future portrayal of them as a "voice" in the design process that Porter (2001) depicts in a contribution to an Association of Teachers of Technical Writing (ATTW) electronic mailing list discussion:

> The political challenge is changing the institutional and disciplinary politics to allow tech communicators a voice in the early phases of the design process. We have to argue for such a position and be able to argue for its benefit. When we adopt a limited role for ourselves—as mere "translators" or transferrers—then we accept the role of waterboy/watergirl rather than player.

I have also observed that technical communication specialists often describe the current work of technical communicators idealistically, perhaps as a way to argue that technical communicators should now (not later) be functioning in more influential roles, as in Bernhardt's (2001) electronic mailing list contribution:

> [Technical communicators are engaged in] knowledge management, leveraging organizational know-how, building expert and memory systems, helping teams work with information, guiding science with documentation, and other such high level activities, while writing manages work, rather than reporting or construing it.

At this stage, most technical communicators lack the power and influence that characterize these high level activities.

To resolve our current identity and credibility crisis, we need to make both external and internal changes. In addition to engaging in external organizational politicking and strategic positioning, we need to modify our internal collective consciousness to leverage our diversity rather than bemoaning that our lack of consensus to date is some kind of tragic flaw that we might never be able to overcome.

Over the past half-century or so, our field has been characterized by an almost incredible array of diverse perspectives. Consider, first, the unusual number of titles of the field that have emerged over the past five decades, all of them characterized by varying frequency of use and concurrence in the field. Quite a few titles have appeared briefly and sporadically, but have remained relatively uncommon, perhaps because they have been used almost exclusively by academics or by practitioners and rarely with equal frequency across those domains. Take, for example, the titles *professional writing, workplace writing, workplace rhetoric, workplace literacy,* and *nonacademic writing* (terms used almost universally just by academics) and *information architecture, information engineering,* and *information development* (terms more common among practitioners).

The most common titles, partly because they have been used with equal frequency in both industry and academia, have been *technical writing, technical communication, document design,* and *information design.* In the 1980s and early 1990s, most specialists in the field abandoned *technical writing* in favor of the more inclusive term, *technical communication,* which reflected the emerging dominance of online documentation in the field and the growing complexity of knowledge required by technical communicators, who, more than ever before, needed to know and do a great deal besides writing. Perhaps because the title *technical communication* is part of the name of the field's largest organization, the Society for Technical Communication, it has remained popular despite the increased use of *document design* in the 1990s and of *information design* in the past few years. Although the title *information design* is gaining wide popularity, it continues to compete with the title *technical communication:* Take, for example, the frequent use of both terms in STC's journal, *Technical Communication,* and in this anthology.

In addition, our field has been unable to reach a consensus about how best to define itself. Classic definitions of technical writing are cataloged and analyzed by Britton (1975), Miller (1979), and Dobrin (1983). Schriver (1997) and Redish (2000) offer more recent definitions of document design and information design, respectively. But no two definitions have been alike, and most of them have included unique features.

Because technical communicators do different things and emphasize different functions, depending on their specialization and organization, no single title reflects everything that we do or wish to do. The range of our functions and roles is unusually large, but this is a healthy reflection of the desirable diversity that characterizes our field. Technical communicators do not even aspire for any single function or role to define the field as a whole.[3]

Why has our field emerged with so many titles, definitions, and perspectives about itself? Possible reasons all reflect healthy characteristics about the field:

• Technical communication specialists have different political agendas, depending on whether they reside in academia, business, or government, and on which organization they serve. In some industry contexts, for example, the title *information design* is likely to elicit more respect than the title *technical communication;* similarly, in some academic English department programs, *technical writing* must be used instead of *technical communication* to avoid the appearance of overlapping with what communication departments teach.

• The field is fluid, not static. The nature and parameters of technical communication, by necessity, must change along with the workplace. Consider, for example, how in the 20th century, the field's title changed from *technical writing* to *technical communication* mostly due to the impact that computer technology had, simultaneously, on businesses as a whole and on technical communication in particular. Similarly, in the 21st century, businesses are now moving into the "information age" and emphasizing the importance of strong design skills, so it is reasonable to expect the emergence of a new title for the field such as *infor-*

---

[3]Although of course, at various times, certain job titles and duties have dominated the field, often to the detriment of technical communicators' status and organizational influence.

mation design. It can be difficult to determine who shapes, manages, and owns information, so now we're competing with other fields with claims to information design expertise, such as *information science, information development,* and *knowledge management*. Information science has even started to place information architects, content strategists, and knowledge managers in jobs that traditionally have been slated for technical communicators. Because our field must change continuously, it might never be possible to develop a permanent, shared perspective of the field.

• The field is context-bound. No two work contexts are alike, so it stands to reason that no two jobs in technical communication are identical. What technical communicators do at one organization likely differs from what they do at other organizations. Similarly, each organization is likely to be unique in the way that it defines and measures best practices or the quality of documentation. As a result, defining what is "best" in technical communication is as challenging as identifying what is common across organizational sites.

• The field is interdisciplinary. Modern technical communication has borrowed considerably from a variety of related fields, including rhetoric, linguistics, graphic design, psychology, organizational communication, and computer science. Because this field is tied so closely to other disciplines, developing a separate identity or definition of its own, as well as determining the field's scope and parameters, is an ongoing challenge.

• The field offers multiple career paths and specializations, which may or may not be considered different from technical communication. As Anschuetz and Rosenbaum describe in chapter 9 in this volume, technical communicators pursue a variety of career paths or specializations (e.g., project, information, or knowledge management; human factors; usability testing). Some specialists in the field consider this a "brain drain," arguing that these technical communicators are leaving the field for specializations. Others prefer a broader perspective, arguing that when technical communicators pursue specializations, they expand their roles while remaining within the field.

My personal background illustrates how both perspectives can be possible. Between the 1970s and the present, I have held a variety of jobs in business, government, and education, with a variety of titles in roughly this chronological order: freelance writer, senior medical writer, editorial assistant, neurosurgical editor, technical writer, researcher, teaching assistant, instructor, assistant professor, senior information design specialist, technical editor, communications analyst, and associate professor. Throughout these two decades, I have always considered myself in the field of technical communication, even though my job titles could easily have masked that continued status. Yet, others could argue that I was in technical communication only when called a *technical writer* or *technical editor* and when teaching courses in technical writing and communication at various universities. As this example illustrates, the parameters of our definitions (and whether they include multiple career paths and specializations) can determine the parameters of our work and field.

The field's context-bound, interdisciplinary, and fluid nature, and its tendency to branch off into multiple career paths, are strengths as well as opportunities for future growth. Consider the unique set of skills and characteristics that technical communicators can contribute to organizations: an ability to understand the particular contextual norms and requirements of an organization; an interdisciplinary, fluid perspective of information development; and the ability to move smoothly between multiple career paths and specialties. The diversity and expansiveness of the technical communicator's skill set and vision can be of great value to any business, and a strength that the field could promote. As technical communicators vie for greater status and prestige within individual companies and find themselves competing with other specialists for control over information design, development, and management, diversity is an advantage that the field can use to gain more respect.

ARTICULATING A PROFESSIONAL VISION

To move the field closer to professionalization, technical communicators need to articulate a clear vision for the field's future. We have made a good start with STC, which has set goals for the field and has established committees to explore the possibility of professionalization. But if STC and other organizations are to take the lead in articulating a vision for the field's future, they need to move beyond the current scope of their mission and goals.

The mission of STC is "to improve the quality and effectiveness of technical communication for audiences worldwide" (*Technical Communication Membership Directory*, 1999, p. 16). This mission statement reflects the overall goal of technical communicators quite nicely, but it was not meant to be a vision statement for the field, nor could it serve as such. What is missing is where the field intends to head and what it intends to accomplish in the new century.

STC's goals are listed here. Notice that goals 2 through 6 focus on internal activities of the organization and what it needs to accomplish to remain productive and self-sustaining. Only the first and final goals address long-term goals that the field could adopt to become more visionary and future-oriented:

1. Enhance the professionalism of the members and the status of the profession.
2. Provide information through publications, reports, and conferences.
3. Report on new communication technologies, methods, and applications.
4. Provide recognition and awards.
5. Provide services to members at all levels of the Society.

6. Promote the education of members and support research activities in the field.

7. Give service to industry and academe.

The first goal touches on some major underlying needs in the field: (a) to improve the status of technical communication as a whole, and (b) to improve the status of individual technical communicators. The seventh and final goal—to give service to industry and academe—mostly refers to STC serving the needs of technical communication specialists in industry and academia, but a broader interpretation could be that it refers to those specialists serving workplace needs or contributing to improvements in workplace contexts. STC goals are not intended to be comprehensive in addressing all major needs of the field. Nor are they meant to bring about major reform in the field or to identify ultimate outcomes (or end results) for technical communication that it hopes to assist or accomplish.

For technical communication to make significant progress forward in the 21st century and to mature into a profession, it needs to formulate and then explore key questions such as these:

• What future do we envision for the field within industry? Ideally, how will technical communicators be positioned in industry, how will they contribute to industry goals, and what will be their job functions and responsibilities?

• What future do we envision for the field across organizational boundaries and in society as a whole? Ideally, how will technical communicators contribute to knowledge about information design that transcends what might be unique to any particular business context? What kinds of questions do they need to help answer, and what kinds of research do they need to conduct to improve the lives of the general public?

• What stake in "the information age" do we envision for technical communicators? How will their work and contributors differ from those in related professions? What will be unique in what we do and in how we can contribute?

• How will technical communicators define, measure, and achieve quality in what they do and in their products? What will academic programs need to do to achieve accreditation, and what will practitioners need to do to earn certification? What changes need to take place in academia and industry to train technical communicators and to evaluate their work? How can measurements of quality be contextualized, so that they can focus evaluation on goals that each organization or organizational culture defines as important?

ORGANIZING PROFESSIONAL VISION AND REFORM

I contend that a new consortium that consists, from the start, of a diverse membership would be an effective mechanism for articulating a vision for the future and bringing about radical change. Before describing this proposed consortium, I discuss why we need this new infusion of organizational energy.

Achieving Consensus: So Far, An Elusive Goal for the Field

A critical first step toward major reform is to find a new, more effective way to achieve consensus in the field about its future. Savage (1999) argues that achieving a consensus about the future is essential for professionalization of the field:

> It does seem significant that at this point the STC is working so hard simply to move toward a more unified self-perception within the profession. Without achieving this goal, it is unlikely that the field can ever realize the advantages of governing itself; it is unlikely that it truly can function as a profession and become recognized socially and politically. (p. 371)

Unfortunately, no consensus is yet in sight about whether the field should move forward to achieve professional status. Carliner (2001) and Davis (2001) both contend that a lack of consensus has been a major obstacle to progress toward professionalization of technical communication. At first glance, because STC has served as such an effective leader of the field for so long, it might seem the logical vehicle for leading technical communication toward reform. Davis (2001), however, points out that even though STC committees have been discussing issues of accrediting academic programs and certifying individual practitioners, they "have not yet suggested workable ways of managing the self-assessment of our profession—perhaps because they have been debating *whether* to do so, not *how* (p. 143)." [italics are hers]

Another recent attempt was made, beyond STC, to reach a consensus about the future of the field. The Summit, a relatively small group of elected organizational leaders and other invited specialists in the field, met fairly regularly in recent years to identify underlying problems in the field and consider and then implement solutions for those problems. The Summit successfully generated a list of problems in the field along with possible solutions for them, but their attempts to help the field achieve major reform are unlikely to succeed fully. Time and resources are scarce for members of this group. Moreover, its membership does not represent the wide diversity of the field, and, as a relatively small and informal group, it has been operating without a broad base of consensus.

I contend that for any organization in technical communication to take the lead in articulating a vision for the future and to bring about major reform in the field, it needs to accomplish the following:

- Acknowledge the need for a broad mission.
- Recognize that important voices in our diverse field have not yet been heard in approaches to date of achieving consensus and making policy.
- Define itself as a coordinator of change; instead of attempting to implement change on its own, it needs to call on diverse talent in the field to do so.

- Ensure that those participating in planning the future of the field really do represent the diversity of the field, and are elected and not self-selected.
- Strive tirelessly for a consensus perspective, without giving up too easily when it appears difficult to achieve that goal.

Of course, organizations that are already formed with structures and agendas different from those suggested here may find it difficult to transform themselves in these ways.

Achieving Reform: What a New Consortium Could Accomplish

A new consortium of organizational members would be ideal to bring about major reform in the field. The new consortium needs to be inclusive, ensuring broad representation of the diversity of the field. It also needs to publicize all aspects of the decision-making process and allow specialists from diverse aspects of the field to participate in generating ideas, voting on decisions, and implementing goals. Membership in this consortium should strive to emulate Davis's (2001) concept of the ideal group to lead the field toward professional status, "a strong, knowledgeable, cross-functional team" (p. 144), including some that do not belong to any professional organization.

To minimize self-selection of membership, consortium members should be nominated and then elected, proportionately, first from major organizations in the field such as the STC (representing mostly practitioners) and the Association of Teachers of Technical Writing (representing mostly academics), and second from a list of academics and practitioners who are unaffiliated with any organizations.

Once the consortium forms and convenes, my recommendations for their central tasks are as follows:

Establish Itself as an Inclusive, Diverse Cross-functional Coalition of Academics and Practitioners.   The coalition should reach consensus, first, about who should lead its efforts.

Articulate a Vision for the Field (in a Vision Statement) and Develop a Comprehensive Set of Goals for the Field's Future.   Most technical communicators could develop an impressive list of goals to accompany a new vision statement. Here is a sample, mostly based on arguments made by contributors in this anthology, that illustrates what this kind of list might look like:

- Elevate the status of the field so that it becomes a profession.
- Modify or elevate the role of technical communicators so that they become leaders in their organization and the value of their work equals that of others high in the hierarchy of their organizations.

• Expand the role of technical communicators so that they can participate actively in product conception, design, and development, along with end-product evaluation and revision.

• Encourage and support more research (by academics, but especially by practitioners to ensure that more research addresses problems identified by practitioners), more quality research, and the dissemination of research results to a wide audience that includes the general public.

• Strengthen the relationship between technical communication and related fields, so that we can coexist, partner, or collaborate with stakeholders in those fields, rather than compete with them.

• Strengthen the relationship between academics and practitioners in the field so that they can find ways to support each other toward separate and mutual goals.

• Allow all voices in the field to be heard and encourage all members of the field to participate actively toward all of these changes.

Create a Strategic Plan and then Follow Up By Coordinating the Implementation of the Goals in Specific Workplaces.   The coalition should develop a strategic plan that outlines goals that the field intends to fulfill over the next 5, 10, 20, or 50 years, and then specific steps for pursuing those goals. Because coalition members are likely busy, they should not be expected to shoulder the burden of implementing goals on their own. However, they should be willing to coordinate the implementation of goals. Their tasks should include establishing and leading committees of diverse members of the profession that can share the work. Implementing the goals should begin in just a few pilot work sites and then be evaluated carefully before they are applied to more work sites.

Create a Set of Standards that Academics and Practitioners Can Strive For and Evaluate Once Technical Communication Achieves Professional Status.   At the pilot work sites, did implementation of individual goals, or of a collective set of goals, work well toward contributing to major reform at the work sites? Do they have the potential to work well at other work sites, or should they be revised first? Should new goals be articulated and tried out? If initial attempts to bring about major reform for the field were less than fully successful, should the consortium try new approaches?

CONCLUSION

This chapter argues for the need to acknowledge that serious problems do exist in the field of technical communication that require major reform. In particular, it calls for elevating the status of technical communicators and of the field as a whole through professionalization. Technical communicators can accomplish only so much individually and internally. Because a lack of consensus has been a chronic obstacle to professionalization of the field, I recommend that a new coalition be established that is inclusive and reflects the diversity of the field. This coalition should have the potential to bring about major structural and perceptual reform and to elevate the field in dynamic ways, so that technical communicators can make significant and valued contributions to work sites and find greater fulfillment in the work they do.

REFERENCES

Bernhardt, S. (2001, March 23). *Association of Teachers of Technical Writing (ATTW) Electronic mailing list* entry.

Britton, W. E. (1975). "What is technical writing? A redefinition." In D. H. Cunningham & H. A. Estrin (Eds.), *The teaching of technical writing* (pp. 9–14). Urbana, IL: National Council of Teachers of English.

Carliner, S. (2001, May). Emerging skills in technical communication: The information designer's place in a new career path for technical communicators. *Technical Communication, 48*(2), 156–167.

Davis, M. T. (2001, May). Shaping the future of our profession. *Technical Communication, 48*(2), 139–144.

Dobrin, D. N. (1983). What's technical about technical writing? In P. V. Anderson, J. Brockman, & C. R. Miller (Eds.), *New essays in technical and scientific communication: Research, theory, practice* (pp. 227–250). Farmingdale, NY: Baywood.

Johnson-Eilola, J. (2001, March, 23, March, 25, & April 4). *Association of Teachers of Technical Writing (ATTW) electronic mailing list* entries.

Miller, C. R. (1979). A humanistic rationale for technical writing. *College English, 40*(6), 610–617.

Porter, J. E. (2001, March 23). *Association of Teachers of Technical Writing (ATTW) electronic mailing list* entry.

Redish, J. C. (2000). What is information design? *Technical Communication 47*(2), 163–166.

Savage, G. J. (1999). The process and prospects for professionalizing technical communication. *Journal of Technical Writing and Communication, 29*(4), 355–381.

Savage, G. J. (2001, April 4). *Association of Teachers of Technical Writing (ATTW) electronic mailing list* entry.

Schriver, K. A. (1997). *Dynamics in document design.* New York: Wiley.

Technical Communication Membership Directory. (1999, September). Special Edition of *Technical Communication, 46*(3-A), 16.

7

Taking Our Stakeholders Seriously: Re-Imagining the Dissemination of Research in Information Design

KAREN SCHRIVER

KSA Document Design & Research

It is an exciting time for research in information design. Researchers are now in the rather luxurious position of being able to draw on nearly a century of findings about rhetoric, writing, visual design, psychology, culture, and human communication. Not only are researchers expanding on past work, but they are also developing hybrid lines of inquiry that cross disciplinary borders and break methodological stereotypes. The last few decades have been marked by growth not only in the number of studies[1] carried out, but also in the forums for sharing this work, for example, journals, electronic mailing lists, and conferences.[2] On the surface, research in information design appears healthy and vigorous.

Scratch the surface, however, and it becomes evident that not everyone would characterize the state of information design research as healthy, especially if positive feedback from the field's many stakeholders is an index of vitality. E-mail posts made to Web-based electronic mailing lists in the field (e.g., TECHWR-L and InfoDesign-Café) suggest that practitioners of information design are negative to lukewarm in their feelings about research in the field. Practicing information designers argue that research is out of touch with the everyday problems of

---

[1]For example, a survey of dissertations between 1989 and 1998 yielded some 200 studies that contribute to knowledge about professional and technical communication (Rainey, 1999).

[2]For a discussion of the development of journals and conferences in the field, see Schriver (1997, chapter 2).

information design. Many contend that the field seems headed in two directions, one toward confronting the communications challenges faced by organizations, the other toward building a research base within academia. Although building a research base is highly valued by academics, practitioners deem the activity irrelevant unless that base connects to their needs.

Practitioners are not the only stakeholders who may feel that information design research has been unresponsive to their needs. Teachers remind us that questions about the pedagogy of writing and design have not been answered by research. Few studies have been carried out in the information design classroom. Teachers confess that for some topics, they must "wing it" because so little is available about what works. At this point, teachers may believe that they will wait a long time before they get research-based advice on, for example, developing students' sensitivity to visual and verbal language. As "insiders" to the field of information design, teachers and practitioners are therefore crucial stakeholders for research.

But the stakeholders for information design research extend beyond insiders, beyond those who already know what the field is about. Citizens and consumers — the major outsider stakeholders for research — are largely unaware of the field. These "ordinary people" do not realize that research could help explain their frustrations with communications that fail them. People are so accustomed to poor communications that it may be hard for them to believe that bad writing and wretched design are not "just the way it is." In fact, citizens and consumers ritually blame themselves when they check the wrong box on a ballot or when things go haywire with their computers (Schriver, 1997). Research that would enable them to recognize their rights for usable information design is rarely made available in forums they read, listen to, or watch. The probability that important stakeholders for knowledge about information design, both inside and outside the field, may feel ignored should concern everyone involved in information design research.

In this chapter, I argue that researchers need to re-imagine the practice of dissemination to help the field take more of its stakeholders seriously. The effect of making research accessible for multiple stakeholders could be dramatic, increasing overall awareness of the value of good information design. More public awareness would likely put pressure on organizations to make excellence in writing and design a priority, a move that would benefit people inside and outside the field.

To show how re-imagining the dissemination of research could prove valuable for the field, I organize this chapter into two parts. The first part characterizes some problems that motivate research in information design. By looking at the problems information design researchers want to solve, one can envision a variety of stakeholders who might find answers to those problems of interest. The second part profiles some of the groups who could benefit from information design research, suggesting that stakeholders differ dramatically in their expectations for research and in their interpretation of what a benefit from research would mean. I conclude by identifying some rhetorical strategies for reaching more stakeholders and suggesting why changing our dissemination practices could benefit the entire field.

PROBLEMS THAT MOTIVATE RESEARCH IN INFORMATION DESIGN

We can learn a lot about a field by looking at the problems it tries to solve and the questions it asks about those problems. In the field of information design, some questions grow from problems faced by practicing information designers. Others emerge as researchers try to understand basic issues of writing, design, and human communication. Still others evolve as communications problems bubble up at work, school, or play. The questions raised by the information design community have tended to focus on features of effective communication processes and products, and increasingly, on the challenges of making communications effective across contexts, cultures, media, formats, and technologies. We can glimpse the field's preoccupations by examining its research agendas— those high level consolidations of the field's unresolved issues.

To identify some issues that have concerned information designers, Fig. 7.1 presents a snapshot of problems that have motivated research in the field. It collapses three research agendas that were composed between 1989 and 2000 (Mehlenbacher, 1997; Schriver, 1989; Society for Technical Communication, 2000).[3] I consolidated the issues raised in these agendas by grouping and re-phrasing them for brevity. As one would expect, the authors of the agendas were not in perfect agreement about what should be studied, but surprisingly, the three agendas clustered around 10 key issues:

1. Developing as an information designer.
2. Anticipating the needs of audiences and stakeholders.
3. Designing information visually.
4. Writing and designing for international use.
5. Teaching information design.
6. Developing methods for people-centered design.
7. Assessing technologies for design.
8. Making decisions about media.
9. Understanding trends affecting the field.
10. Criticizing and refining ideas about research.

---

[3]The first was generated by me in 1989. The second agenda, from 1997, was created by Bradley Mehlenbacher of North Carolina State University. The third, from 2000, was compiled by the Research Grants Committee of the Society for Technical Communication, a committee that funds proposals for research.

1. Developing as an information designer

- How do information designers develop the knowledge, skills, and sensitivities needed to excel in writing and design?
- What constitutes expertise in information design? What are the characteristics of information designers who are experts?
- What is the role of information designers' knowledge? Subject-matter knowledge? Linguistic knowledge? Perceptual knowledge? Rhetorical knowledge? Procedural knowledge?
- How does multidisciplinary collaboration shape the nature of the information design process? How should information designers prepare themselves to work on virtual teams?
- In what ways can information designers influence product and service design as well as information design?

2. Anticipating the needs of audiences and stakeholders

- How can information designers better meet the needs and expectations of diverse audiences (e.g., those who differ in age, gender, ethnicity, knowledge, skill, experience, or motivation)?
- How do the needs of experienced audiences differ from those of lay audiences?
- What strategies are effective for anticipating and meeting the needs of multiple audiences and stakeholders?
- How can information designers meet the cognitive and affective needs of audiences with special requirements, such as children, older adults, and people with limited technical proficiency or language skills?
- How does the communication medium influence stakeholders' thinking and actions as they engage with text and graphics?

3. Designing information visually

- What are the principles of good visual design?
- How can information designers make better decisions about when to present information visually, verbally, or in some combination?
- What are effective ways to integrate visual and verbal information (in hardcopy, online, or multimedia formats)?
- Do stakeholders in different fields have unique expectations for texts and graphics?

4. Writing and designing for international use

- How can information designers develop a better understanding of cultural differences and their implications for writing and design?
- What are the features of a good plan in designing information artifacts for international use?
- What are the best ways to evaluate the adequacy of visual or verbal information for international use?
- What features of the context need the most attention when tailoring information to the unique needs of a local community?
- What strategies should information designers consider in avoiding cultural bias in their writing and design?

5. Teaching information design

- What do information designers learn from observing their stakeholders, users, or audiences interacting with text and graphics? Are there long-term benefits?
  What is learned from usability testing and can it be consolidated and taught directly?
- What are the most effective methods for teaching information design—that is, for teaching writing, design, and the integration of the visual and verbal?
- What teaching strategies work best in the university classroom, onsite in organizations, or in distance-learning environments?

FIG. 7.1. Key ideas from research agendas composed between 1989 and 2000 (Mehlenbacher, 1997; Schriver, 1989; STC, 2000).

6. Developing methods for people-centered design

- How can information designers refine their ideas about people-centered design and usability?
- Which evaluation methods are best suited for judging the quality of paper-based artifacts, Web designs, or information displayed on a palm-sized screen?
- Can information designers develop more sensitive evaluation methods than are currently available? What are the effective combinations of existing methods?
- What makes electronic information (e.g., embedded help, Web, and wireless displays) easy to use, memorable, and fun?
- What planning and management methods best support the design of usable and maintainable information?

7. Assessing technologies for design

- What kinds of technologies best facilitate writing, design, production, and evaluation?
- What constraints does technology place on the information design process? Are some technologies counterproductive for good work?
- How does GroupWare enhance or impede collaboration and information development?
- How will the convergence of communication technologies (such as the Web, TV, telephone, and video) affect the possibilities for information design?
- What is the relationship among publishing systems, tools for collaboration, and online information delivery programs? How do these relations change in distributed environments and those that use object-orientation (e.g., SGML, HTML, XML, and VRML)?

8. Making decisions about media

- What is involved in developing effective communications for paper, electronic formats, or the Web?
- What are the factors in selecting and combining communication media, such as paper, online, and multimedia?
- How should information designers make choices about the distribution of content across media (e.g., what kinds of content are better presented on paper versus online)?
- What are the best strategies for information designs that serve multiple functions, for example, to inform, teach, and persuade?

9. Understanding trends affecting the field

- How do trends influence information design (e.g., trends in society, politics, business, and technology areas)?
- How will the field of information design change as a result of organizations that are increasingly global?
- Will the trend toward wireless and mobile computing change the types of information designs that will be needed?
- What are the implications of single-sourcing for multiple-use information designs?
- How do information designers acquire an understanding of best practices in the field?

10. Criticizing and refining ideas about research

- How can information designers bridge the gap between the knowledge generated by practitioners who work in corporations and teachers and researchers who work in universities?
- How can information design practitioners, teachers, and researchers better share their work with each other and with other constituencies?
- Can information designers develop alternative models for disseminating research? Are there hybrid genres that might reach wider audiences and stakeholder groups?

FIG. 7.1. (*continued*)

On the Dissemination of Information Design Research

Researchers whose studies have shed light on the issues presented in Fig. 7.1 have—for the most part—disseminated their work through books and articles. Most of these publications have been directed toward an academic audience. As one might expect, researchers usually disseminate their work to insiders of the field before outsiders. In the course of dissemination, research results are typically presented first to "exclusive insider groups" such as other researchers. As the work gains momentum, ideas about it may or may not be diffused to insider groups other than researchers, for example, to practitioner communities. Only rarely, if the work has enough traction and enough street interest, do the findings eventually reach stakeholders outside the field via forums such as newspapers, television, or e-lists. The process is long and circuitous, but eventually the work could get out to insider and outsider stakeholder groups.

For information design, a less-than-well-established field with few publishing venues, one might expect the dissemination process to be fraught with slowdowns, stagnation, and stalls. And it is. But one would not necessarily expect the process to grind to a halt before the work is disseminated to the variety of publics it may interest. Unfortunately, the dissemination of information design research has tended to stop once it reaches the academic journal, technical report, or book. This has made it close to impossible for people outside of the field to gain access to work in information design. Alas, much of the best scholarship of the 1980s and 1990s was published in obscure journals, books without indexes, technical report series (now unavailable), or anthologies that rapidly went out of print. After publication, researchers tended to move on to the next project rather than refashioning their findings and ideas for diverse groups of insiders and outsiders. I contend this has been a mistake. This unwitting insularity has prevented the field from fully realizing itself.

Part of the problem is that researchers in and out of academia have been rather parochial in how they imagine making their work visible. As a consequence, many dissemination efforts culminate in a single publication to an audience much like themselves. Researchers rarely reach what is perhaps the largest potential audience for their work: practitioners and teachers in their own field. This is unfortunate because, as Fig. 7.1 shows, researchers have been interested in questions about design processes and products—questions practitioners and teachers wrestle with every day. Clearly, practitioners and teachers have been top "interested groups" for research-based ideas.

For example, practitioners on the job often need data to bolster their arguments to managers "without a clue" about what matters. That so many organizations fail to understand the value of information design is a serious and widespread problem for the field generally, affecting members in both academic and nonacademic positions (Ramey, 1995; Redish, 1995; Schriver, 1993).

Main (2001) critiques what might be called "the cult of *RoboHELP*"—that is, the pervasive practice of employers who use knowledge of *RoboOffice 9* as the prerequisite for employment. He encourages employers to look beyond a person's ability to master a new authoring tool (e.g., *RoboHELP, FrameMaker, HomeSite*) and to examine actual expertise involved in information design.

Clearly, employers need better information about what expertise in writing and design actually involves and the role that technologies may play in enabling writers and designers to do their best work. They need to know when technology may facilitate production, collaboration, visualization, or maintenance. They also need to know when technology enables thinking and when it just gets in the way and hobbles creativity. Well-disseminated research could go a long way toward changing employers' ideas about what is important in hiring information designers.

Of course, employers would not be the only beneficiaries of research on expertise in information design. If teachers of information design had better data about the knowledge, skills, and sensitivities that information designers need, they could design their courses to convey a more vivid picture of what students will face as new professionals. They could be more creative in their selection of course content and better able to develop content that challenges students' abilities to think visually and verbally.

Unfortunately, practitioners and teachers are frequently forgotten when researchers disseminate their work. An even bigger challenge for researchers, however, lies in reaching the many stakeholders outside the field. Results about what works would attract the attention of stakeholders from the general public, especially those who find themselves needing to make design decisions without any experience. Professionals who write or design "by default" would find research useful that prevents them from making poor decisions.

Of course, the general public might especially value ideas about how to avoid getting hoodwinked by information design. They would also appreciate research that helps them sort out good information design from bad. For example, consider the now infamous butterfly ballot from the year 2000 presidential election, an information design nightmare that captured public attention around the world. It was redesigned (actually, tinkered with) by a well-meaning but misguided election official who made design changes based only on her intuitions about what would work. She should have known about usability testing, but she did not. If voters had known about usability testing, they might have realized that their tacit approval of the ballot (after glancing it over when it appeared in local newspapers) was a mistake. They would have insisted on a process allowing people like themselves to try it out first, instead of just looking it over. Just one short hour of usability testing could have avoided the weeks of counting and recounting in the butterfly ballot fiasco. As this example illustrates, consumers can be helped significantly when they understand the impact of design in their lives.

Taking the 10 issues presented in Fig. 7.1 together, it is clear that the field of information design has raised some crucial questions that if answered, even partially, would benefit a variety of stakeholders. Some of these stakeholders would be curious about the findings for practical reasons. Others would be concerned with how the findings challenge previous theory and research. Still others would value any new ideas about writing and design. That research in information design only occasionally reaches these varied stakeholders is worrisome.

STAKEHOLDERS FOR INFORMATION DESIGN RESEARCH

In this part, I explore expectations that stakeholders for information design research may hold. As I have suggested, researchers have not done very well in meeting some stakeholders' expectations. However, it seems likely that researchers' failure to disseminate their work broadly has not been out of malice, selfishness, or arrogance. Rather, it may be that researchers have had trouble finding the time to sufficiently consider the range of stakeholders for their work and their possible interests. My intent here is to flesh out some ideas that researchers might consider in making these early but crucial decisions about making their work visible.

Researchers need to consider the relationship between themselves and the possible stakeholders for their work. If no relationship exists, they need to consider how they will connect with stakeholders. This requires taking a hard look at the research from the stakeholder's point of view. In some cases, stakeholders are much like the researcher, expecting a technical report of research. But more often than not, stakeholders are unlike the researcher in at least four ways:[4]

- *Expectations* for the genre, media, and format in which they would prefer to hear about the research.
- *Biases* for the tone of the communication, especially about the "certainty" of the conclusions; for example, whether the author hedges the claims or presents them forcefully as definitive.
- *Assumptions* about what an interesting research finding would be, that is, different assumptions for what constitutes a "good" result.
- *Strategies* for reading, watching, listening to, scanning, and searching through communications about research.

Understanding these differences—*expectations, biases, assumptions,* and *strategies*—is crucial in planning communications for diverse stakeholders. One way

---

[4]I present these four ways as hypotheses about differences among stakeholders for research. They are derived from my previous research in information design (Schriver, 1997), particularly my work in observing how people interact with documents.

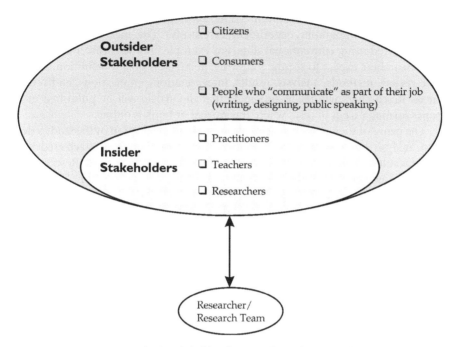

FIG. 7.2. Multiple stakeholders for research in information design.

to anticipate the needs of constituencies such as teachers, practitioners, or the general public is to assess what each group knows about information design. If people are generally knowledgeable, they are probably insiders. If they are unaware of information design, they are likely outsiders to the field. For convenience of discussion, Fig. 7.2 shows the distance between the researcher (or research team) and possible insider and outsider stakeholders for the work.

Who Are Our Stakeholders and What Do They Want?

To provide a more specific idea of what may be involved in trying to reach the stakeholder groups shown in Fig. 7.2, I suggest next what people from each group might expect as they engage with research about information design. I speculate on likely expectations, biases, assumptions, and strategies that each group may bring to what researchers typically conceive of as a "report of research." I offer these characterizations as hypotheses to be tested.

Insiders to the Field: Researchers

Expectations for Genre, Media, Format. As ultimate insiders to research, other researchers will expect an article-length report that will follow the genre conventions and formats (e.g., APA, MLA[5]) typically associated with the study's method (e.g., ethnography, experiment, case study). This means the article will be structured using conventional slots. For example, in a scientific journal article, researchers expect to see the following: abstract, problem definition, literature review, methods, analysis, results, interpretation, implications, and references. Researchers will likely anticipate that the article will be published in a paper format, except in cases where the journal or book is online.

The genre of the research article calls for careful explication of the study's design, analysis, and results. In fact, whether other researchers even attach credence to the results depends on their evaluation of the design and analysis. Researchers implicitly agree on the "rules" for making good arguments (Abelson, 1995) and understand there is a rhetoric of empirical research (Hayes et al., 1992). Researchers respond more positively to studies that err on the side of modesty than those that overclaim or fail to hedge about the certainty of a finding.

Biases for Preferred Tone and Certainty of Conclusions.   Researchers tend to be scrupulous in their concern about the relationship between the data collected and the claims made. They feel they know a good design when they see one. They look for a fit between the data, methods, and analysis. In experimental studies in which inferential statistics are reported and the results fail to reach statistical significance, researchers expect rhetorical hedging of the conclusions and implications. In a qualitative study, they scrutinize the relationship between the claims made and the data, asking, "Do these data add up in the way the researcher suggests?" Too much certainty in stating qualitative conclusions may be viewed as arrogance or naivete.

Assumptions About What Constitutes an Interesting Finding.   Researchers agree that "good data" are the best evidence in support of an argument. Data do not speak for themselves;[6] their selection and display are rhetorical acts. Data must be interpreted with rigor and honesty. Conclusions need not be hyped in order to be interesting. An interesting result either (a) challenges or confirms previous theory and research, (b) lays a foundation for new lines of inquiry, or (c) surprises and shocks because it confronts widely held beliefs about the issue under study.

Strategies for Reading, Watching, Listening, Searching, and Scanning.   Experienced researchers tend to read the article's title, abstract, data displays, conclusions, and references (see Berkenkotter & Huckin, 1995; Gross, 1990). Early in their scanning of the piece, researchers attempt to identify the conversation the

---

[5]APA refers to the stylistic conventions of the American Psychological Association; MLA refers to the conventions of the Modern Languages Association.
[6]See Abelson (1995).

research speaks to and speculate on the likely novelty and interestingness of the study. Researchers may verify the author's understanding of the conversation by checking the references (note: this checking is often done immediately after reading the article's title instead of after reading the conclusions). If, after a brief scan, the article looks sufficiently interesting, the researcher will likely go back and read the whole piece with care.

Insiders to the Field: Teachers

Expectations for Genre, Media, Format.   Teachers expect that the article will give them a sense that the researcher has actually been in a classroom.[7] They prefer articles that employ a conversational style. Teachers tend to dislike articles that follow the standard genre conventions for reports of research. Case studies, narratives, and personal testimonials are highly valued. Teachers expect the article will be presented on paper but may also be available in a variety of electronic formats (e.g., Web or CD-ROM).

Biases for Preferred Tone and Certainty of Conclusions.   Teachers who sense that researchers have not experienced the dynamics of a classroom will view the work with suspicion. Similarly, researchers who describe students as faceless "subjects" will be met with skepticism. Teachers are alienated by researchers who overgeneralize or who offer a one-size-fits-all approach to instructional strategy. Not surprisingly, teachers focus their reading on the practical ideas about what does or does not work in the classroom. Teachers expect "take-home messages" to be presented separately from the data, which they would prefer not to read.

Assumptions About What Constitutes an Interesting Finding.   Teachers expect that the article will do more than "just lay out a problem"; it will give research-based advice. The article will interpret data in ways that ask teachers to reflect on their methods, strategies, and "ways of doing" in the classroom. A good result is one that challenges teachers to rethink their current practices.

Strategies for Reading, Watching, Listening, Searching, and Scanning.   Teachers tend to read the title, skim the article for main points (e.g., what takes place in the article), and hunt for anything related to "implications for teaching." If the implications appear relevant, they may go back and read the whole piece.

---

[7]I base my description of teachers on 10 years of experience working as Research Associate for the Center for the Study of Writing, a national center that was shared between the University of California at Berkeley and Carnegie Mellon University. During the 10 years of its existence (1985–1995), the Center for the Study of Writing had a mandate from its funder, the U.S. government, to disseminate its work to high school and college teachers. Part of my work for the Center was to act as publications coordinator for our dissemination efforts, a task that proved difficult because we researchers struggled with finding a voice that would speak to teachers.

Insiders to the Field: Practitioners

Expectations for Genre, Media, Format.   Practitioners expect the article to "tell the story" of the research—an enlivening narrative about the problem under study and the implications of the findings.[8] The article need not follow conventions for research reports; a magazine style format is preferred (see, for example, the format of *Intercom*, the magazine of STC). Practitioners hope the article will blend narrative and workplace examples. Like other stakeholder groups, practitioners still like their paper documents, but unlike some other groups, they also expect delivery in a variety of well-designed electronic formats.

Biases for Preferred Tone and Certainty of Conclusions.   The writing and graphics will reveal that the author has experienced what it is like to be a writer, designer, or user. Practitioners expect the article to highlight problems faced on the job, such as working under severe time and financial constraints. Practitioners usually prefer a conversational to slightly formal tone. They like graphical displays of conclusions and want the argument to be straightforward, devoid of subtlety and nuance. Practitioners hope the piece will emphasize the implications of the work, that is, what they can do with the results. It will de-emphasize theory, research methods, data collection, and analysis. Practitioners look for conclusions phrased as guidelines in their most general and action-oriented form.

Assumptions About What Constitutes an Interesting Finding.   For many practitioners, an interesting finding sheds light on a difficult problem and offers practical ideas for solving it. A good article helps practitioners (a) increase their knowledge, (b) develop data-based challenges to other professionals who often hold power within organizations (such as managers, programmers, and engineers), and (c) rethink their practices (including technologies for practice).

Strategies for Reading, Watching, Listening, Searching, and Scanning.   Practitioners tend to be attracted to distinctive titles. If a title sounds interesting, they read the abstract and the pullout quotes. It is likely that they then scan the visuals, the displays of data, and the itemized lists. Practitioners may then go to the conclusions section and to see if there are any useful guidelines or principles. At this point, practitioners make a judgment about the potential value of the article. If deemed promising, they go back and read the whole thing.

---

[8]I base my description of practitioner preferences based on working with Maurice Martin, editor of STC's magazine *Intercom*. For several years while editing a column, *Research in Technical Communication*, I received a great deal of feedback about what practitioners like and do not like and offer these ideas as summaries of what I heard.

Outsiders to the Field: People Who "Communicate on-the-Job"

Expectations for Genre, Media, Format.   People who might read informative articles about "communication on-the-job" expect the sort of provocative thought pieces found in airline magazines or the illustrative case studies of the *Harvard Business Review*.[9] These readers like to hear about the everyday struggles faced by people who are not very good at writing and design but who are nonetheless called on to write or design themselves or to manage people who do so. These busy people are often managers (or managers in training). In my experience, they often like to hear about new ideas outside of their domain via radio, television, or viral marketing (ideas spread person-to-person) before they are willing to invest time reading about that idea. Fortunately, many people who need information on-the-job recognize the potential value of checking Web sites on the topic. This adds another communication channel for information design researchers.

Biases for Preferred Tone and Certainty of Conclusions.   Some professionals prefer to read articles that adopt the informal tone of a colleague who offers advice. Other professionals, particularly those in technical, scientific, or medical industries, may be more comfortable with the formal tone of a technical report. Some are suspicious of "friendly documents" because they are bombarded with online and hardcopy sales literature that aims to be friendly. Most professionals look for elements of a human-interest story along with elements of a white paper (e.g., engaging narrative supported with high-level summaries of data). Professionals who are accustomed to reading trade magazines tend to look for cases or vignettes that illustrate key findings.

Professionals without a background in research tend not to hold the same standards for rigor as professionals with training in research. Most professionals are not particularly bothered by data that is inconclusive, even though they are more persuaded by data-based claims. Professionals tend to listen more carefully to the findings of researchers and organizations they believe have reputations for being good. Appropriate or not, professionals make assumptions about the quality of the thinking based on the credentials and affiliations of the researcher. Thus, most professionals would give more weight to a story with the headline: "Researcher from Stanford Finds that Editing Online Leads People to Miss Problems in their Writing . . ." than the more ambiguous headline: "Professor Finds that Editing Online Leads People to Miss Problems in their Writing . . ."

Assumptions About What Constitutes an Interesting Finding.   For many professionals, an interesting finding offers practical advice about a familiar problem,

---

[9] I base my descriptions in this section on 20 years of consulting experience. In seminars on effective communication for managers, I often asked participants about what they expected from a written piece that would help them on the job.

for example, tips on how to achieve a credible persona in Web design. These readers want ideas for improving their writing and design in tangible ways. People on the job often find interesting ideas about the design of common organizational genres (e.g., office correspondence, customer letters, reports, brochures, newsletters, house organs, Web sites, and the like).

Strategies for Reading, Watching, Listening, Searching, and Scanning.   People on the job are usually willing to read an engaging thought piece—that is, as long as the piece is not too long, no longer than say two to four pages. People on the job are often unwilling to read lengthy material online because so many sit in front of a computer for more hours per day than they would like. These stakeholders would prefer to watch or listen to the message rather than read it. Still, many of them look for hardcopy so they can show the piece to colleagues on the job.

Outsiders to the Field: Consumers and Citizens

Expectations for Genre, Media, Format.   Consumers and citizens—the largest of the potential stakeholder groups for information design research—share many expectations. They expect communications that are brief and informative and that offer "guidelines." One model for consumer- or citizen-oriented guidelines can be found in the public service announcements from Science in the Public Interest. Researchers from this consumer watchdog group told the public about the high fat content of American Chinese restaurant food and offered guidelines for consumers about ordering dinner in Chinese restaurants. They used plain language and simple quantitative graphics to explain the issues.

Although they may not be able to put it into words, most consumers hope their perspective will guide the structure of the communication. Readers who are unfamiliar with a procedure feel more comfortable, for instance, with the use of scenario headings in which human agency is put into focus (e.g., "What every consumer should know about emergency evacuation procedures").

Consumers and citizens want to know what research about information design is good for and why they should be paying attention. They respond favorably to reports that rattle common assumptions and myths (e.g., "Information Design Researcher Finds that Women Are Just as Good at Programming Their VCRs as Men, Even 70-year Old Women!"). Consumers and citizens look for useful information in many venues—newspapers, magazines, ads, TV spots, radio announcements, tabloids (*The National Enquirer*), mass-market magazines (e.g., *Readers' Digest* or *Consumer Reports*), flyers, brochures, posters, catalogs, and Web sites.

Biases for Preferred Tone and Certainty of Conclusions.   Not surprisingly, the tone that consumers and citizens prefer is conversational, the style plain and direct. An article about information design research could reach a wide audience if it were a human-interest story supported by surprising data. The data, presented in their most condensed form (e.g., one sentence and/or one visual), might be framed by a vivid case or a startling finding. For better or worse, consumers and citizens tend to focus on the interestingness of the case and/or findings and may not consider whether the data are conclusive. It is common for people to look only at the "take home" message that might be meaningful for them. And not surprisingly, many consumers and citizens would prefer to listen to radio or watch TV in order to learn about research.

Assumptions About What Constitutes an Interesting Finding. A good result informs consumers, often by confirming their suspicions (e.g., manuals for digital cameras are hard to understand) or by encouraging them to recognize their rights (e.g., even fine print should be intelligible). Good research helps consumers see familiar situations, as Proust said, with new eyes. A useful study gives citizens language for "talking to each other" about a communication issue. It also gives them evidence for "talking back" to organizations that provide inadequate communications (e.g., some computer software makers, consumer electronics companies, financial firms, pharmaceutical companies, and government agencies). Moreover, good communications about research puts citizens in a more informed position—helping them play a role in shaping future communications from business and government. By raising citizens' expectations for communications, researchers can raise the bar for quality in information design within organizations.

Rhetorical Strategies for Reaching a Wider Group of Stakeholders

We can see that these various stakeholder groups—researchers, teachers, practitioners, workers, consumers, and citizens—might differ considerably in their expectations for a "report of research." These divergent needs place significant demands on researchers, calling on them to articulate clearly what it is that others may find engaging about their work. Researchers need to specify who might find their work (or some part of it) of value. Once researchers have a good understanding of what stakeholders might find interesting, useful, or valuable, they can employ rhetorical strategies for shaping their communication more directly.

As information design researchers know perhaps too well, stakeholders are an impatient lot. They expect to gain an immediate grasp of answers to questions such as: Why is this interesting? What practical use is this? What value will this have for me? When information is presented in ways that violate people's expectations, they tend to give up on the communication. To avoid alienating stakeholders before they fully engage with the report of research, researchers need to think rhetorically about how to present their work. Here are some strategies that might be considered at the beginning of the dissemination process:

Consider the Options for Genre(s).   In some cases, the traditional report of research is exactly what is called for. However, in others, it would be more appropriate to choose a well-defined genre other than the research report (e.g., the magazine article). In still others, it would be most effective to combine genres in novel ways (e.g., a full-page newspaper ad with persuasive appeals about a topic that also includes major elements of a one-page research brief).

Adapt the Content and Structure of the Report to Connect With Stakeholders' Prior Knowledge and Beliefs.   Learning is more effective when we build on what people already know about the subject, even if what they know is wrong (Ausubel, 1963). To effectively adapt the content of a research study to the needs of a stakeholder group, researchers need to conduct a thorough audience analysis. Information about what stakeholders know and value about the topic is crucial for building conceptual frameworks to guide further understanding and appreciation of the topic. By strongly connecting with peoples' prior knowledge through the macrostructure of the text (e.g., devices such as headings, subheadings, topic sentences, photos, graphics, and data displays), the researcher can cue the reader that "this is interesting, keep looking." Many information designs fail to engage because nothing strikes the reader as visually or verbally intriguing.

Stakeholders' prior knowledge can also be acknowledged through well-chosen examples. It is often useful to include examples from the data that tap into stakeholders' values and misconceptions about the topic. Realizing stakeholders' values through text and graphics can maximize the chance that people will relate to the content.

Present the Text and Graphics in a Level of Detail and Complexity That Suits Stakeholders' Interests.   Presenting information at an appropriate level of detail and complexity typically calls on researchers to abandon the stylistic conventions of academic discourse (e.g., pompous diction and heavily embedded syntax). Instead, they need to employ the conventions of journalism (simple words with easily parsed subject–verb–object structures).

Attending to the level of detail necessarily involves attending to length. A difficult aspect of meeting stakeholders' needs lies in finding ways to discuss complex research topics briefly. Most potential stakeholders for research do not want to hear everything about the work. In most cases, they want only the main point. For them, the shorter the better. Even so, creating short versions of research may challenge researchers because they become attached to their content in all of its glorious detail. However, too much detail may make it hard for some readers to get the main point; it will make others just stop reading.

Reconsider the Media for Presenting or Displaying the "Report."   In many cases, researchers need to explore the use of media other than paper. This may mean creating an oral summary for radio, but more often, it involves generating a shorter and more graphically sophisticated version that can be displayed online or streamed on the Web. This requires reframing the piece to make it more modular, more visual, more like spoken language.

Although there is nothing surprising about these principles for people educated in rhetoric, it is surprising that researchers tend not to draw on their rhetorical training when disseminating their work.

TWO CASES IN POINT

In this chapter, I have characterized why it is important for researchers in information design to re-imagine the dissemination of their work. I have sketched out ways in which the field's research agendas may hold value for diverse stakeholder groups and posited what different stakeholders might expect from a "report of research." I have also offered strategies that researchers might consider in reaching these different stakeholders:

- Rethink the options for genre(s).
- Adapt the content and structure to connect with stakeholders' prior knowledge and beliefs.
- Present the text and graphics in a level of detail and complexity that suits stakeholders' interests.
- Reconsider the media for presenting or displaying the communication.

To illustrate these ideas, Appendixes A1, A2, B1, and B2 provide examples from two research projects. These examples show how each research project was reconceived for different audiences.

A Case Study of Usability Research

See Appendixes A1 and A2. The first example is from a usability research project my colleagues and I did for a large Japanese consumer electronics company. In the study, we evaluated the quality of a set of instruction guides for a variety of consumer products (VCRs, TVs, and high-end stereo systems). Our research team revised the guides on the basis of extensive usability testing and expert review. We consolidated what we learned from our research in the form of guidelines and principles. At the end of the project, we sent a summary of our work to the company in the form of a traditional technical report, the first page of which appears in Appendix A1. The audience was our client: Mitsubishi Electric's marketing department. The marketing group was enthusiastic about the findings and about the possibility of improving their products but also spreading the word about their efforts.

Not long after submitting the report, an Associated Press reporter, Jeffrey Bair, interviewed me and wrote a newspaper article about our team's work. His article was printed by newspapers that use the Associated Press (AP) and Knight Ridder wire services. Trade magazines within the consumer electronics industry also picked up Bair's story and elaborated on it with interviews with Mitsubishi management. Appendix A2 presents one of these magazine-style features. A third, TV version of the story (not shown) appeared on CNN's *Science and Technology Today*. The segment dramatized the idea of wretched usability by having one of my students use a set of "read and weep" instructions for a high-end musical digital synthesizer, the sort used by Led Zeppelin (but not easily).

The important thing to notice about this example is how Appendices A1 and A2 differ in their organization and selection of content. The technical report starts with a conventional "executive summary," whereas the magazine piece jumps into a humorous setup. The technical report has a formal tone and is focused on "the facts"; the magazine piece is conversational and oriented toward human interests. The report must account for what happened, that is, tell the story of the project; the magazine article motivates readers to understand why the story is important to them.

A Case Study of Research on Drug Education Literature

See Appendixes B1 and B2. Here, I present a snapshot of a research project about understanding how teenagers interpret drug education literature. In this study, my colleagues and I explored how teenagers responded to the writing and design of brochures that were meant to encourage them to "just *say no* to drugs." Over several years, my colleagues and I talked with hundreds of teenagers about their interpretations of these brochures. At the end of the study, I wrote up the findings for a book chapter. Appendix B1 presents the first page of the study as it appeared in the book, which was directed toward advanced practitioners, teachers, and researchers of document design (see Schriver, 1997).

In contrast, Appendix B2 presents a more popular depiction of the drug education research: a press release. Although Appendixes B1 and B2 have almost identical titles, the selection of content and the organization are quite different. The book chapter takes the perspective of problems faced by professionals who design drug education literature. The press release takes the taxpayer's perspective and implies that citizens are footing the bill for drug prevention programs that may not be working. (The book chapter eventually makes a similar claim, but not nearly so quickly and simply.) The book chapter assumes the audience wants a story replete with interactions, conflicts, and complexities. The press release takes for granted that the reader wants pithy examples of why the drug education materials are not working. The story was the same, but was addressed to different stakeholders with different values for engaging with the content.

These two cases show how a single body of research can be refashioned to meet the needs of different audiences and stakeholders. Naturally, this kind of reworking of original material takes time. It involves difficult reorganization and slashing of content. But the time researchers invest can pay off as they reach new audiences who come to know that information design research is responsive to real human needs. By re-imagining dissemination to include a more diverse group of stakeholders, researchers can energize their work and re-invigorate the field itself.

REFERENCES

Abelson, R. P. (1995). *Statistics as principled argument.* Hillsdale, NJ: Lawrence Erlbaum Associates.

Ausubel, D. (1963). *Educational psychology: A cognitive view.* New York: Holt, Rinehart & Winston.

Berkendotter, C., & Huckin, T. N. (1995). *Genre knowledge in disciplinary communication: Cognition/culture/power.* Hillsdale, NJ: Lawrence Erlbaum Associates.

Gross, A. G. (1990). *The rhetoric of science.* Cambridge, MA: Harvard University Press.

Hayes, J. R., Young, R. E., Matchett, M. L., McCaffrey, M., Cochran, C., & Hajduk, T. (Eds.). (1992). *Reading empirical research studies: The rhetoric of research.* Hillsdale, NJ: Lawrence Erlbaum Associates.

Main, M. D. (2001). The undervaluation of writing expertise. *Intercom, 48*(3), 12–14.

Mehlenbacher, B. (1997). *Trends in information design.* Keynote presentation to the SAS Institute, Inc., Research Triangle Park, NC.

Rainey, K. T. (1999, November). Doctoral research in technical, scientific, and business communication, 1989–1998. *Technical Communication, 46*(4), 501–531.

Ramey, J. (1995). What technical communicators think about measuring value added: Report on a questionnaire. *Technical Communication, 42*(1), 40–51.

Redish, J. C. (1995). Adding value as a professional technical communicator. *Technical Communication, 42*(1), 27–39.

Schriver, K. A. (1989). Document design from 1980 to 1989: Challenges that remain. *Technical Communication, 40*(2), 239–257.

Schriver, K. A. (1993). Quality in document design: Issues and controversies. *Technical Communication, 40*(2), 239–257.

Schriver, K. A. (1997). *Dynamics in document design: Creating texts for readers.* New York: John Wiley and Sons.

Society for Technical Communication. (2000, November). A research agenda proposed for 2000. *Intercom, 47*(9), 25.

APPENDIX A1

An original technical report of research. The stakeholders for the work are high-level managers of the marketing department of a Japanese consumer electronics company.

How Consumers Understand Mitsubishi's VCRs, TVs, Stereos: A Usability Evaluation

Document Design Team:

Karen Schriver, Research Director	Don Kline
Christopher Freeble	Michele Matchett
Nancy Downes	Joy Jones
Diane Haugen	Deborah Ainsworth

Communications Design Center
Carnegie Mellon University
April 4, 1990

Executive Summary

This report presents the results of a two-year project for Mitsubishi Electric Sales of America (MESA) conducted by the Communications Design Center of Carnegie Mellon University. The project involved evaluating the writing and design of instruction guides for VCRs, TVs, and stereos. It began with a features analysis of the guides to assess the quality of the originals. This expert evaluation was followed by usability testing—an assessment of the guides and the product interface by members of the intended consumer market for each product line. Usability testing was followed by a wholesale revision of each guide. These initial revisions were then put through an exhaustive technical review with experts in audio/video systems. Based on input from these experts, the guides were again revised to generate a prototype for each product line. Finally, the lessons learned were consolidated into a set of guidelines for writers, designers, and technical illustrators. These guidelines cover writing, design, typography, interface design, logo design, and technical illustration. Mitsubishi can now build on the results of this project to demonstrate how it is making its products easy to use and easy to learn. Moreover, Mitsubishi now has a platform for future document design and interface design.

Introduction

A few years ago Mitsubishi began to notice that customers were complaining about how difficult it was to use its products. People were unable to program their VCRs to record, use the menus displayed on their high-end TVs, or hook-up their surround-sound stereo systems. Although Mitsubishi was producing outstanding technologies, it was having difficulties communicating effectively with its customers.

Mitsubishi studied the product design, the interface, and the instructions to locate the problem. All three were found to cause users troubles. To solve the "instruction manual" problem, MESA switched to videos. Instructional videos were created and shipped with all new VCRs. Unfortunately, customers who needed help "setting up their VCR" could not play the tape until it was set up, rendering the tape effectively useless at the time customers most needed help. Marc Auerbach of MESA recommended going back to paper, but this time to "get it right." Subsequently Mitsubishi approached Karen Schriver to review the manuals.

APPENDIX A2

A magazine article based on a newspaper report created by Jeffrey Bair of the Associated Press. Bair's original article appeared in over thirty newspapers around the world. This adaptation (without an attribution of authorship) appeared in the glossy trade piece "Consumer Electronics Society News" in Spring 1990.

MIRROR 6 **SPRING 1990**

New Owners Manuals designed to eliminate blinking "12:00"

It's one of Johnny Carson's favorite jokes: The clock on his VCR is forever locked on "12:00," its blue display blinking into eternity. He can't figure out how to program the machine and he can't decipher the owner's manual.

"Twelve. Twelve. Twelve," says Oprah Winfrey. "It drives me nuts. And the most frustrating thing is that sometimes the (owner's) manual directions are more confusing than just looking at the pictures."

No one seems to enjoy reading owner's manuals and it doesn't help when they're hard to understand. They often look like they were written by engineers as an afterthought. After all, not many consumers base their purchasing decisions on the owner's manual, so manufacturers have had trouble justifying the resources required to develop well-written, easy-to-read owner's manuals.

Well, the Mitsubishi Electric Sales of America (MESA) Audio/Video Group (AVG) is out to change all

It's one of Johnny Carson's favorite jokes. The clock on his VCR is forever locked on "12:00," its blue display blinking into eternity. He can't figure out how to program the machine and he can't decipher the owner's manual.

that. Nearly two years ago, the AVG commissioned Carnegie Mellon University's Communications Design Center in Pittsburgh to rewrite and redesign owner's manuals for MESA's entire line of televisions, videocassette recorders, and audio systems.

"We're trying to do something to simplify the lives of the people who buy our equipment," says Leo Delaney, Vice President, AV Marketing.

The new manuals are written in plain English. Jargon is virtually eliminated. Active voice

replaces passive voice. Illustrations are simple and easy to understand. Liberal use of white space keeps the material uncluttered. A detailed index and an exhaustive table of contents help users find information quickly. A glossary explains technical terms used in the text. Headers and footers identify chapter topics at a glance to make skimming through the manuals easy.

MESA may be the first company to include the design of the owner's manuals as part of its marketing efforts. "We think it makes a lot of sense to let people know we are doing

everything we can to help them understand their equipment," Delaney says.

Redesigning owner's manuals isn't as easy as it sounds, according to Chris Freeble, AVG's Communications Design Specialist. Freeble was part of the original team that worked on the MESA project at Carnegie Mellon. After graduating with a master's degree in professional writing, he joined the AVG to put the research into practice.

Carnegie Mellon has an entire department devoted to the study of document design. The field encompasses cognitive psychology, graphic design, human factors, and professional writing.

Under the direction of Dr. Karen A. Schriver, a group of graduate students identified problems with the owner's guides, constructed guidelines for designing the manuals and applied the new rules.

The new manuals will be phased in gradually, beginning this spring, and Delaney expects to have them completed for the entire A/V line by the end of the year.

APPENDIX B1

An excerpt from a discussion of the research that appeared in my book *Dynamics in Document Design* (1997, pp. 167–168). The book was directed toward an audience of professionals, teachers, and researchers.

"JUST SAY NO TO DRUGS" AND OTHER UNWELCOME ADVICE: TEENS SPEAK OUT

Recently my colleagues and I[21] studied a context in which good writing and visual design have the potential to make an important difference: the design of drug education literature. We were concerned with how teenage audiences interpret brochures intended to discourage them from taking drugs, and more broadly with how readers may respond to the visual and verbal messages presented through brochures that aim to inform and persuade. We felt that the area of drug education literature would provide a challenging rhetorical situation to study because it is a context in which the audience's knowledge and values may stand in stark contrast to those of professionals employed to write and visualize the documents. Professionals who design drug education literature typically differ from their audiences in age, in point of view, in experience with drugs, in education, and sometimes in race, culture, and social class. Designing documents that communicate across these social and cultural boundaries is complex because professionals may have difficulty in anticipating how someone who may be quite unlike themselves will interpret their ideas.

Furthermore, even when professionals are good at "getting on a level" with their readers, the organization sponsoring the document may constrain the "voice" document designers can create by controlling (and in the worst cases, censoring) what may be said or illustrated.[22] This study showed us how critical it is to consider the possible interactions and conflicts among the values of the document designer, the organization, the gatekeepers, and the intended audience. It also made us aware of how important it is to learn about what audiences believe and value by listening to them as they interpret documents.

Where Our Research Team Started

We began by collecting over 100 brochures and handouts from national and local drug prevention agencies.[23] Many of these materials were funded by U.S. taxpayer dollars or through grants to nonprofit organizations during the Reagan administration.

APPENDIX B2

A press release based on the research on drug education literature described in Appendix B1. The audience is the general public who may read newspaper articles.

Karen Schriver Associates, Inc.   33 Potomac Street
Document Design and Research   Oakmont, PA 15139
                                (412) 828-8791
                                (412) 828-7247 FAX
                                schriver@cmu.edu

NEWS FOR IMMEDIATE RELEASE

FOR FURTHER INFORMATION:
Karen Schriver, Ph.D.
412 828-8791

"Just Say No To Drugs" and Other Unwelcome Advice: Teens Speak Back!

> *"These people must think we are really stupid."*
> —A 7th grader from Pittsburgh

PITTSBURGH, PA—Feb. 27, 1997—Research on public information about drug prevention finds that brochures designed to encourage teens to "Just Say No to Drugs" may evoke scorn and ridicule. A study by Dr. Karen A. Schriver in *Dynamics in Document Design* (Wiley, 1997) explores how teenagers react to the words and pictures of anti-drug brochures. Teenagers in the study found that drug prevention materials were often "condescending" and "corny." For example, one brochure produced by Health and Human Services during the Reagan-Bush era recommended turning down offers of drugs by saying, "No thanks, I'd rather walk my pet python." Teenagers tended to characterize the writers of the brochures as "seriously out of touch." One 8th grader offered this impression: "The writer sounds like somebody who would never come to my neighborhood, but who wants to control us." Said another, "It seems like it was written by someone who sits in an office all day and gets all their information from books." This research suggests that taxpayer dollars are sometimes wasted on drug prevention messages that backfire with teenagers—showing the crucial role of testing what works and what doesn't.

—more—

8

Migrations: Strategic Thinking About the Future(s) of Technical Communication

BRENTON FABER
JOHNDAN JOHNSON-EILOLA
Clarkson University

Clarkson University seems to be an appropriate place for the series of discussions we have had that led to the creation of this chapter. Located in St. Lawrence County in upstate New York, we are about a 45-minute drive to the Canadian border, about 90 minutes from Canada's capital city, Ottawa, and nearly the same distance from the metropolitan center of Montreal. An hour south are Lake Placid and Adirondack National Park, which still holds the lodges of some of America's wealthiest families. The park has recently enjoyed a renaissance of tourism, hosting numerous business conferences, two international triathlons, a marathon, numerous hockey tournaments, and figure skating competitions. Last spring, bird watchers from across the world came to the park to see the three-toed woodpecker, a rare species in North America.

Clarkson's faculty is comprised of researchers from nearly every continent on the globe, and the school is a world leader in numerous scientific and engineering fields. Like most universities, we enjoy the latest technological innovations. Our students come from all over the world, and many bring with them even more expensive automobiles than those located in the business school parking lot. The local college town, Potsdam, New York, boasts Mexican, Indian, Chinese, and Italian restaurants, and until her recent move to Switzerland, Canadian country music star Shania Twain was occasionally sighted at the local grocery store.

At the same time, "The North Country" continues to be the poorest region in New York State. St. Lawrence County, where Clarkson in located, is the third poorest county in the north. Per capita income in St. Lawrence County in 1996 was $15, 994. Per capita income in the North Country was $17,110. In 1990, single parents headed 14.3% of the county's households, and 47% of these families had annual incomes below the federal poverty line. Between 1980 and 1990, the number of mobile homes in the county increased by 62%.

Thus, in many ways, Clarkson embodies the best and the worst of globalization. On the one hand, globalization has created a diverse and advantaged community in this relatively remote outpost. At the same time, unemployment in this region has remained stubbornly high, rural poverty is endemic, few people actually born in the area ever graduate from one of the three universities here, and fewer university graduates stay in the area after graduation. The gap between "town" and "gown" continues to grow as the gap between rich and poor, cultured and illiterate, mobile and trapped also widens.

Our purpose in this chapter is not to present key ways this globalization is changing patterns of work and economic value and how these changes may influence the ways technical communicators are positioned in the workplace and in the economy. We do not suggest that unless technical communication as a field embraces globalization, we will end up on the wrong side of the globalization duality. Such an argument would be callous and ill-conceived. It would also work against much of what we are currently doing to try to obviate the many problems associated with globalization in our community. Nor is this chapter a critique of globalization. We view globalization as a complex and diverse process that does not lead to obvious answers. Instead, we hope that this chapter prompts a larger discussion within the field of technical communication of organizational value, knowledge value versus product value, and the global effects of our work on both growing and struggling economies.

Globalization, as described by Micklethwait and Wooldridge (2000), columnists for *The Economist,* refers to the integration of the world economy into one large market (p. xvi). According to this model, products will have a global audience and distribution, and financial decisions will respond to and influence economic conditions far beyond any single country's borders. As demonstrated by our brief introduction to the economic conditions surrounding Potsdam, globalization is commonly seen to bring with it great financial advantages for some people, but tremendous burdens and hardships for others. Increasingly, these gains and losses are the result of two fundamental changes in modern economic markets: the decline of labor value and the accompanying rise of intellectual capital. This shift in value has occurred because in a global context, industrial work can be outsourced to the country or the community that offers the lowest wage and production cost, thereby devaluing the process of production. Whereas this process was typically associated with low-skilled labor, in a global context, even high-skilled labor can be exported and sold to the lowest bidder.

As outsourcing drives down the value of production, products can no longer compete solely on price. They must compete on design, function, usability, and the value they add to the consumer. In such a context, what becomes valued is not the actual product, or the ability to make that product, but the ability to imagine, design, innovate, and teach new products and new methods of production. In the context of global competition, the ability to create and access new knowledge, share that knowledge throughout the company, and then leverage that knowledge into new products and services becomes more valuable than the ability to simply manufacture a product. Whereas production is the key feature of the industrial economy, knowledge is the key feature of the information economy. In this discussion we elaborate on this distinction and demonstrate how it applies to the ways technical communicators position themselves within organizations.

This switch in the ways the economy derives value has several troubling implications for technical communication. As a practitioner-based field, much of our knowledge creation and use has been directed toward the development and creation of products. More problematic is that often our products simply support other products. Thus, we have defined our value through the products technical communicators produce: manuals, help files, interfaces, Web pages, courses, and tutorials.

Even though the transition to an information-based economy has certainly spurred overall growth in technical communication and improved the overall situation of most academics and practitioners in our field, in this chapter, we argue that this growth is expectant but not sustainable. What we mean here is that people outside our field have predicted the potential value that technical communication as a knowledge-based resource can bring to a project. However, because the field itself has been less willing to take on the role of knowledge producers rather than product producers, this current growth trend will not continue interminably.

In a broad and abstract sense, if technical communication is to thrive in an information-age economy, our field as a whole must develop an entirely new way of understanding the relations between school and work and between knowledge production and knowledge use. To develop this new understanding and take on more powerful and valued roles in the emerging knowledge economy, we need concretely to strengthen the ties between academia and industry.

Although our field has debated the nature of the relationship between industry and academy for decades, we have done little in a systematic way to build better relationships based on knowledge creation rather than on process efficiency or product creation. Our field's prejudice toward product knowledge and away from creative or innovative knowledge building is evident in the many practitioner calls we have heard for academic research to recommend the most efficient font types, color coding schemes, or usability tests, rather than calls for more innovative ways for technical communicators to build the knowledge base of an organization.

Our argument in this chapter focuses on what we call a corporate–university hybrid. This hybrid involves mutually supporting institutions that together enable each other to be stronger players in a knowledge economy. This hybrid counters the tendency to position academics and practitioners in theory/practice, knowledge/product, or creator/user dualities and the tendency to view professionals in each world as competitors for scarce resources. Instead, the hybrid fosters a view of academics and practitioners as currently occupying distinct and separate professions tied by a common insight into the value communication brings to organizations and workplace processes. By working together to reinforce our strengths and common visions, we can provide valuable learning and workplace experiences for future technical communicators in both industry and academia.

THE PROBLEM: EDUCATION AND WORK IN THE AGE OF INFORMATION ECONOMIES

The perceived gulf between academics and professionals comes down, in the end, to this: they are two sides of the same information economy coin (or e-coin™). However, like other theory/practice tensions that were created to support and advance the needs of industrial production, our current academic and corporate structures are designed to ensure a constant supply of well-trained workers who can support the creation and maintenance of physical commodities. Similarly, academic research in this industrial model may suggest new efficiencies, model new approaches to product building, or critique existing systems. By remaining segmented from practice, however, theory remains sufficiently removed from the production process to not interfere with practitioners' basic aptitudes and capabilities to get their work done. Industrialism requires basic research to incrementally improve efficiency and skills. However, an industrial economy cannot sustain continual or wholesale changes over an extended period of time. Thus, any research that is not incremental or supportive of the basic pragmatic needs of production is considered out-of-touch, abstract, unproductive, or in other words, theoretical.

As a field that has emerged from, and continues to depend on, industrial relations, technical communication is not well-suited to developing the constantly changing and increasingly complex capabilities demanded of knowledge workers. Because academics and practitioners in technical communication largely share this industrial model of education, research, and work, both groups tend to assume that the spheres of education and work are primarily separate.

For example, at the Milwaukee Symposium, a recent gathering of academic and industry representatives who debated the future of technical communication, participants dedicated a relatively large amount of time to discussing ways in which academics could provide practicing technical communicators with research that was easily applied to concrete workplace tasks. Again, academics were expected to provide incremental, product-based knowledge that resulted in increased workplace efficiencies. Missing from this conversation was a discussion of whether or not such information was particularly valuable. What we mean here is that any advice provided about the rules for concrete tasks—font selection for a particular Web page, for example—was not going to be easily extensible to other similar situations (because of the interactions between font size and color space, monitor size, and line length, let alone differences among users and specific contexts of use). Although in an industrial context such questions are important because they can increase process efficiencies, the rapid change in technologies and uses of technology makes the current shelf life of such rules of thumb very short. As a result, although such information may end in short-term problem solving, in the long term such task-specific work has little value because it does not help academics or practitioners do their job better (Carliner, 1996; Selber, 1995). From our perspective, what was missing in this discussion was a knowledge-based method in which experts themselves could identify and choose particular tools that were most appropriate for specific problems. The emphasis here, however, is not on the tools but on the knowledge and value-added expert stance such knowledge can create.

The futility of prioritizing technical knowledge is underscored by Reich's (2001) recent revision of his concept of "symbolic–analytic work" from his earlier model (Reich, 1991). He has shifted the focus away from mastery of technology and information and toward the production of knowledge for particular groups of users. Contrary to his earlier definitions, Reich's 2001 work argues that technical skills should not, in themselves, be taken as a primary trait for the new classes of workers he describes. In fact, a deep focus on developing and manipulating technologies draws away from a person's ability to both understand and construct creative solutions for other users. In addition, such a focus returns workers to an industrial emphasis on simply building products with the assumption that products themselves carry value. However, in an age when products themselves have no stand-alone value, what is valued is the ability to see how a specific product can meet a specific user's needs. Building a yin/yang set of qualifications, Reich (2001) applies two stereotypes to his new worker: geek/shrink. The geek half possesses technical ability while the shrink half understands the needs of users and, not incidentally, market-specific methods for leveraging knowledge into consumer–user value (pp. 51–57).

What should strike technical communicators about this information-age geek/shrink model is how closely these stereotypes match up to many of the debates technical communication has had about our professional identity: our debates over the importance of tools and technical knowledge, our insistence on understanding users in real contexts, and our ongoing discussions of how better to market how important technical communication—not simply technical knowledge—is to our society.

However, technical communication, with its continued emphasis on products, is not yet capable of addressing in a systematic way the question of our collective identity. As a field, technical communication remains remarkably fragmented (Ecker, 1995; Jones, 1995; Savage, 1999). Practitioners would be hard pressed to identify a robust core body of knowledge that a practicing technical communicator should possess (witness the ongoing debates over certification). Academics would likewise have difficulty agreeing on a body of theoretical knowledge and research (at either the graduate or undergraduate level) that all students of technical communication should learn. In addition, we remain troubled by the persistent lack of a central societal need or prevalent social cause that defines technical communication in the way that health care (eradication of disease) informs medicine or the pursuit of justice informs law. This lack of a professional basis for the field (a core body of knowledge used to solve an important social need; Friedson, 1996) only exacerbates our focus on building good and better products without a framework that details why and for whom these products are valuable. We are not claiming that technical communication products are not valuable. Instead, we are claiming that technical communication has not built value into its workplace or its academic practices. In many ways, we remain victims of Plato's critique of rhetoric: We do not control how our products are used and as a consequence, we do not control how value is assigned to our work. Unlike knowledge workers, who bring specific solutions to specific problems, we are teaching students to build products for an unidentified and often unknown audience. Thus, our workplace models seem closer to mass-market consumer goods than the high technology information economy with which we have aligned ourselves.

The complex array of skills and abilities required for new forms of work defy traditional approaches to both education and training. Current job tracks for technical communicators tend to rely on two traditional approaches: bottom up, on-the-job learning (perhaps augmented by technical skills or education previously attained) or college or university degrees followed by additional on-the-job training. Although many technical communicators certainly practice many of the skills involved in that job classification, they also frequently tend toward other, less valued job classifications identified by Reich—routine production and in-person services. Although we have begun to move away from forms of technical communication that relegate our work to the routine production of texts according to simple templates and forms, that is the professional image of us that a substantial number of people have in both the general public and the workplace. Perhaps more distressing is the chance that technical communicators become defined as in-person service workers, with their clients being inside experts: programmers, engineers, scientists, and managers. As Carliner (1996) observes, "Because we overemphasize the role of tools in our work, we get pegged as tool jockeys rather than communicators—asked only to convert a file from Word to RoboHELP rather than asked to write the text in the file" (p. 273). Following the economic dictates of globalization, this sort of service position with its continued emphasis on the product—we might increasingly include tasks such as designing Web sites or translating technical documents among other standard technical communication services—is increasingly being supplanted by software that is less expensive, more available, and arguably more efficient than a group of technical communicators. A software-created Web site may not be as fully competent as one designed by a technical communicator, but that argument did not stop companies from building tax preparation software, Travelocity.com, or the automated assembly line pictured in Ford Focus advertisements. But our task is not to make clear why a group of technical communicators can build a better Web site. Our task is to add value to that software by teaching people and companies how to use it to solve their problems. We must find ways to leverage our knowledge and build new knowledge to create and add value in a business culture that is increasingly agnostic to physical products.

Unfortunately, neither current academic programs nor existing professional practices in technical communication are in a strong position to create and support those sorts of leaders. Currently, we rely on academic educational institutions to provide broad, conceptual knowledge—such as audience analysis, graphic design concepts, methodology, and critical "outside the box" thinking. Although certainly students in academia also frequently apply their knowledge, in general we expect their most relevant application to come when they enter the workplace. It is here (according to this career model) that students become actual technical communicators, putting their somewhat abstract theories to work in the "real world" in a relatively linear model of career maturation. The foundation academia builds eventually is capped and in some ways replaced by the bricks and mortar of industry experience.

As Savage notes (1999), it is not clear that this model ever worked well in the first place. It certainly will not serve to push technical communication toward the types of abilities required to excel at symbolic–analytic work in a global information economy. This context requires technical communicators who are constantly reunderstanding and re-presenting their own value in both conceptual and applied ways. It is not a matter, for example, of being able to quickly augment one's own knowledge of HTML with XML or even of understanding the basics of effective interface design. Nor is it a matter of providing an evolving set of conceptual frameworks in the education of technical communicators. These are still product-focused applications. Instead, the rapid pace of change in both applied and conceptual knowledge in global information economies requires a hybrid sort of learning. What we need here is a view of career learning that is integrated in all aspects of professional work, a context in which the value a technical communicator adds is an ability to constantly learn in each situation and integrate that learning into audience-specific solutions.

Consider the rapid adoption of a technology such as Macromedia Flash in professional Web site design. Flash provides designers with tools for developing relatively fast-loading animations on Web sites. The technology itself represents a relatively short learning curve (just hours to learn the basics for someone already experienced with other applications). However, as with other hybrid skills, using Flash to design effective, usable Web sites requires much more than understanding how to place objects in motion on the screen. Whereas many academic programs in technical communication provide graduates with a practical framework for understanding effective interface design in terms of static graphic elements, the addition of temporal factors to the interface changes the situation dramatically. For example, technical communicators typically learn to draw out tree or network structures to represent a complex text. Such approaches are appropriate for structurally static texts, but less suitable for texts that shift over time. Programs offering animation or dynamic activities often center design activity about a layered timeline. Although the mechanical shift from one representation scheme to another can be achieved relatively simply, the shift in conceptual framework and design practice is much more profound.

This is not to say that practicing technical communicators will not, over time, develop rules of thumb for designing effective Web animations (in fact, many technical communicators already have developed such skills) based on work tasking, the accumulation of scores of hours of practice, observation, and feedback. But such on-the-job learning fails to provide an effective framework for (a) rapid development of rules for use, (b) the ability to test and revise hypotheses about use in sound ways, and (c) the development of a strategic understanding of the software as a value-added business solution. To respond to new technological uses—not merely new pieces of software, but new ways of working and communicating—in ways that are both fast and effective, technical communicators must routinely engage in work at both applied and theoretical levels. If technical communicators hope to influence their (or another) company's strategic missions, they must leverage their knowledge and skills to add value to the company through new knowledge which, in turn, creates new revenue-generating products or processes. Such work would position technical communicators somewhat differently—as knowledge workers or even as consultants within their own organizations.

SOLUTIONS: INCREASING MIGRATIONS
BETWEEN ACADEME AND WORKPLACE

In the second half of this chapter, we propose a hybrid form of technical communication that would move the field closer to professional status. It is tied to recent trends to partially collapse academic/corporate divisions as exemplified in the development of corporate universities and industry-sponsored research centers in academia. Drawing on these examples, we outline one step that could be fruitful for developing technical communication as a profession suited to advanced symbolic–analytic work in the global information economy. This is work that requires the ability not merely to learn and teach new technologies, but also to provide creative and innovative solutions to unique and specialized business problems.

To provide technical communication in general with robust skills and a professional framework (both theory and practice), we need to increase the migration of information and people between the two contexts. People and ideas move back and forth between industry and academy, supporting each other and constantly mutating and bringing value to both locations. Only this sort of movement will give us mutual respect and, more importantly, the robust framework that we need to develop a connected body of theory and practice. Our current inability to make a strong value-added claim to organizations and our relatively low status within both academic and practitioner organizations may be tied to our failures to unite those two (Savage, 1999).

We already have several models for the starting point of such collaboration. Corporate universities, for example, increasingly provide education to employees within the workplace (or other sites). Unlike traditional training programs, leading programs integrate theoretical problem solving, contextual perspectives of business problems, and a strong pedagogical commitment to a reciprocal relationship between learning and doing. For example, Accenture offers business skills courses that fully integrate management theory, organizational behavior, and research methodology with classroom simulations, problem-based learning, and other forms of learning-by-doing. Other corporate universities, for example, Symbol University, through its partnership with Long Island University, offers Symbol employees MBA programs on-site.

At the same time, academic programs in science, business, and engineering increasingly collaborate with corporate partners to identify and solve both applied and basic research issues. Technical communication programs that exist in schools of engineering may be best equipped at training the symbolic–analytic workers discussed here, provided their work is not subsumed to an in-person or routine-productive service model discussed earlier.

Returning to the context we used to introduce this chapter, Clarkson University, like most academic institutions, hosts a number of corporate/academic hybrids. The Eastman Kodak Center for Excellence in Communication, for example, which Johndan directs, provides faculty and students with an environment in which to engage in both traditional (academic) and project-based learning. This summer, Brenton is working with GE Supply and Accenture to build a course based on corporate models of delivery he observed while researching corporate universities. Students from disciplines across campus currently use online space to show business clients their work. Clarkson faculty members Stephen Doheny-Farina and Dan Dullea are coordinating a student team to produce patient-education videos for the Canton-Potsdam Hospital. In addition to the Kodak Center, the School of Engineering and the School of Business operate several corporate-funded centers, including a Global Supply Chain Management Center and an Internet Consulting Group. These centers have become important hybrid locations where business and academia meet to solve persistent problems and invent new knowledge.

Although many academics (and probably not a few professionals) might discourage such a strong corporate presence on university campuses, the development of hybrid research sites and projects is crucial to the future of technical communication. Without this transformation, technical communication will remain fragmented in the ways we have already discussed. Such fragmentation means that we will continue to be incapable of leading change, or of responding actively and in rich ways to new technological, social, and political situations, and other people will claim the value for our work.

We propose, then, a radically different sort of career path and institution, one that not only allows but also encourages employees to move among educational and corporate institutions, as needed. We are not calling for a collapse of both institutions into a single hybrid entity, but rather attempting to construct much stronger and frequent interaction between the two. This interaction will build on strengths but will also retain the integrity of both sites.

Although obviously there has been substantial interaction between academics and practicing technical communicators for the entire history of technical communication, such interaction has frequently taken place at the margins, in unofficial and unsupported ways. In the most successful cases, individual schools and corporations have formed working relationships, featuring such activities as onsite research in corporations, adjunct faculty from corporate sites who teach occasional courses, and corporate advisory boards. The Society for Technical Communication (STC), for example, offers a faculty internship scholarship designed to supplement salaries to provide academics with job-related experience.

Table 8.1 portrays how current academic/industry interests have aligned in areas of curriculum, products, cross-training, and cross-delivery. Here, the uni-

TABLE 8.1

Comparison of Current and Proposed Models of University/Industry
Responsibility and Cooperation

	Current Areas of Responsibility and Cooperation		
	University	*Joint*	*Corporate*
Curriculum	core knowledge	elective interests	specific training
Products	research, open and free dissemination	sponsored projects	marketable products, proprietary knowledge, restricted dissemination
Cross training	corporate interns	co-developed projects	faculty interns
Cross delivery	graduates	user tests/beta tests	research funding

versity produces theoretical knowledge that creates a core curriculum. This information is freely disseminated through publication and conference presentations. Cross-initiatives are limited to elective interests, interns working temporarily in industry, and graduates moving into corporate settings. Corporate sites offer specific training and create marketable products. Companies also provide opportunities for faculty interns and increasingly they provide research funding for academic projects. At the same time, companies often restrict access to proprietary information and corporate knowledge and the dissemination of that knowledge.

We do not want to underestimate the value of such linkages: They provide important sites for exchange and learning. However, in order to develop a profession that provides the sorts of abilities required in the new workplace, technical communication must reinvent itself in ways that better support the free flow of information and learning between the two.

The process of reinvention will not be easy or simple. Institutional inertia in both academia and industry discourages radical rethinking of their respective roles and structures. In addition, practitioners need to cope with corporate tendencies and policies to restrict access to knowledge and keep proprietary control over their workplace information. Such restrictions to the free dissemination of information seriously limit the ways academic faculty can and wish to participate in collaborative projects.

Table 8.2 compares the current model with a proposed new arrangement between academics and practitioners.

This restructuring emphasizes the importance of knowledge flow between academic and corporate sites. It also emphasizes the move to knowledge-based work in corporate settings. Rather than view corporate settings as the location of product creation, they become the place where experts use new and existing products to solve social and business problems. Products are co-developed

TABLE 8.2
Proposed Models of University/Industry Responsibility and Cooperation

	Potential Areas of Responsibility and Cooperation		
	University	*Joint*	*Corporate*
Curriculum	core knowledge	product specific training	value-added behaviors and knowledge
Products	research open and free dissemination	marketable products	problem solutions based on expert knowledge
Cross training	corporate interns	co-developed curriculum	faculty interns
Cross delivery	research expertise	user tests/beta tests	access to research sites and support of faculty research

between academic research centers and corporate experts. A significant condition for this work is the free distribution of information across academic and corporate sites. Such exchanges will do much to enable faculty research and theory to be more relevant to workplace settings. In addition, such knowledge exchanges will help build a culture of knowledge work in both organizations.

We suggest that both academy and industry radically increase the amount and frequency with which technical communicators move back and forth between the two types of institutions. This activity would include practicing technical communicators returning to the academy to either teach or take a course in a theoretical area (e.g., visual theory, methodology), as well as academics entering the workplace in short-term positions to work on applied issues such as conducting usability tests or interviewing subject-matter experts to draft a feasibility study, or teaching or taking a course at the workplace.

Such migration occurs frequently, as we acknowledged earlier, and usually to great benefit. It is also clear, however, that such migration is not rewarded or supported at the disciplinary, organizational, or even cultural level for most academics or practitioners. For technical communicators to move freely between the two types of institutions, both arenas need to provide real support for that movement—not merely at the level of allowing a practicing technical communicator to teach a course at the local college in her spare time or even at the level of encouraging faculty to pursue corporate funding for research. Real change and real support require positive measures and actual cultural change.

The immediate benefits of such migratory paths are clear: improved, value-added experiences and practices in both academia and the workplace. At a second, broader level, these movements would provide technical communicators with the skills, background, and maturity needed to occupy more central roles in an emerging global economy. We are not asking that academics simply obey the requests of industry or that employers hire students regardless of abilities. Instead, increased circulation between the two contexts would provide academics with better understanding of the goals and constraints of workplace technical communicators. This understanding is critical to build more robust theoretical frameworks that could support practitioners as knowledge workers and creative solution providers—as key members of the new economy. At the same time, increased circulation between the two contexts would give practitioners a better understanding of the broad uses of theory—beyond simple skills acquisition or information delivery.

To conclude, we return to our local context and briefly present a concrete example of what we mean. Information Solutions (IS) is a nonprofit organization in downtown Potsdam that teaches computer-related job skills to people who have suffered traumatic brain injury. These individuals learn basic computer tasks while completing data entry, burning CDs, and doing other word processing and information recording for clients. This past term, Brenton's "Professional Writing" class worked with IS to solve a number of communication and business issues. For example, one student group built an interactive Web page that allowed IS to advertise its business and solicit information from potential clients. Another group wrote a grant proposal. A third group wrote training manuals for clients and staff.

For the most part, the class was a success. Like most service learning courses, students learned about working with an actual client and adapting to actual deadlines and workplace constraints, and Brenton was able to integrate into the course discussions of nonprofit organizations, their role in the economy and society, and different organizational issues nonprofits face. However, according to the terms we outlined in this paper, the class could also be considered too industrial or product-based. The students' focus was on completing their specific products for IS. Students did not consider finding ways IS could leverage its existing knowledge into new products or processes, nor did they recommend different knowledge-based business models IS could use to supplement their cash-starved charity.

IS is one of the few business-based workplace training centers for people who have suffered traumatic brain injury. Students researched and found only two other similar organizations in America. Yet, 5.3 million U.S. citizens (nearly 2% of the population) live with the effects of brain trauma injuries. There appears to be a significant market here for a knowledge-based consultancy that could help other communities and agencies build their own workplace training centers for other people who have suffered brain injuries. This is the kind of value-added thinking we recommend in this chapter, as it creates a synergy among technical communication products and new business models and processes. It requires technical communicators to think beyond products and efficiencies, to engage in actual knowledge work. Such work puts technical communicators in the center of business practices. They choose, create, and deploy products, inventing and directing how these products will work together.

To come full circle, note that our ongoing work with IS is but a small intervention in the way the global economy has affected Potsdam. The people IS serves are among those who Micklethwait and Wooldridge (2000) describe as losers in the global economy. There are few jobs in our area for physically challenged individuals. There are no work-related educational programs and the only options for these people are welfare, social assistance, and other dependency programs. By teaching a variant of technical communication skills and supplying knowledge value, we hope to help IS become a more sustainable organization, while also creating new roles for technical communicators.

REFERENCES

Carliner, S. (1996, August). Evolution–revolution: Toward a strategic perception of technical communication. *Technical Communication, 43*(3), 266–276.

Ecker, P. S. (1995, November). Why define technical communication at all? *Technical Communication, 42*(4), 570–571.

Friedson, E. (1996). *Professional powers: A study in the institutionalization of formal knowledge.* Chicago: University of Chicago Press.

Jones, D. (2000). A question of identity. *Technical Communication, 42*(4), 567–569.

Micklethwait, J., & Wooldridge, A. (2000). *A future perfect: The essentials of globalization.* New York: Crown Business.

Reich, R. B. (2001). *The future of success.* New York: Alfred A. Knopf.

Reich, R. B. (1991). *The work of nations: Preparing ourselves for 21st-century capitalism.* New York: Alfred A. Knopf.

Savage, G. J. (1999). The process and prospects for professionalizing technical communication. *Journal of Technical Writing and Communication, 29,* 355–381.

Selber, S. A. (1995). Beyond skill building: Challenges facing technical communication teachers in the computer age. *Technical Communication Quarterly, 3*(4), 365–390.

Expanding Roles
for Technical Communicators

LORI ANSCHUETZ
STEPHANIE ROSENBAUM
Tec-Ed, Inc.

Thousands of people receive degrees in engineering every year. Ten years later, some of them will still be practicing engineering and designing computer chips, aerospace electronics, or automotive emissions systems. Others, however, will have used their engineering training as a pathway to careers in marketing, product/project management, corporate management, management consulting, engineering consulting, and more. These people have not forsaken engineering. Rather, they have built on their engineering training and experience, embracing related skills and gaining positions that allow them to exert more influence in their organizations, exercise more creativity, or have more choices for professional growth.

The same process frequently happens in technical communication. We are technical communicators who have become practitioners in human factors/usability. Our own experience led us to think about our colleagues with similar experiences, and to sample what we can learn from them about how our profession is evolving.

This chapter provides findings from interviews we conducted with nine professionals in the computing industry, including brief case histories for six of them. Although all have their roots in technical communication, they have expanded their careers by adding a professional specialty such as usability, marketing, information architecture, or project or program management. Our interviewees have not stopped being technical communicators. Rather, they have made career transitions that have given them the opportunity to apply their communication skills and knowledge in inventive ways to new venues. In doing so, they have attained new spheres of influence with respect to product quality and product-line direction.

What does this kind of transition say about technical communication? It suggests that we are at a crossroads, and that choices we make now will greatly influence the future of the field. One choice is to define technical communication traditionally, as a field that involves the design and development of information products—print documentation, online help, interfaces, multimedia presentations—and whose aim is to communicate technologies so that users can assimilate them into their everyday goals and work. An alternative choice is to define technical communication more expansively, as a comprehensive network of activities, knowledge, and skills that help technologies be useful, usable, learnable, enjoyable, memorable, marketable, competitive, and of high quality. In this view, technical communication embraces all of the following efforts:

- Building organizations that value user success with their products and systems, because they believe satisfied customers increase profits.
- Including user experience considerations in strategic planning and corporate decision-making for company and product-line direction.
- Collecting user data and applying it to the scope, design, development, and marketing of products and systems.
- Improving product user interfaces and user support (on-screen and hardcopy) by applying communication expertise.

We propose that this second definition be the one that characterizes our field. Although today's technical communicators rarely perform the functions described in all four points in the preceding list, opportunities exist to move into these roles. Increasingly, technical communicators are moving from the traditional fourth point to the other three, and they are doing so with great success.

The transition to expanded roles often involves new job titles that do not immediately bring to mind traditional technical communication activities. However, title changes do not mean that individuals in these new roles cannot still feel a kinship with technical communication in spirit, allegiance, and perspective. They can, especially if technical communication defines itself expansively.

The case histories that appear next illustrate many ways in which technical communication has integrated with—and prepared for—related career paths such as knowledge management, human factors, information architecture, software development, marketing, corporate management, and management consulting. Thereafter, we explore what these case histories tell us about how the technical communication profession can embrace and support transitions—and ultimately expand career options within the field.

CASE HISTORIES

Most of the interviewees in the case histories have responsibility of one kind or another for collecting user data and applying it to the design, development, and marketing of products and systems. In this capacity, they do more than design and develop information. A few of them also are responsible for bringing user experience to bear on the strategic planning and corporate decision making that determines product-line direction.

In terms of general characteristics, all but one of our interviewees are female. All of them have been working since 1985, many longer; consequently, the cases may represent opportunities and trends related largely to that time period. All of the interviewees have experienced several career transitions in their professional lives. Only one has an advanced degree in technical communication; almost all of them hold undergraduate degrees in fields other than technical communication.

In reporting the interviews, we do not link the interviewees' career transitions to the characteristics of the organizations in which they work—size, profitability, age/maturity, geographic location, industry segment, or mission. Discussions of organizational psychology are also beyond the scope of this chapter. With these qualifications in mind, the case histories are intended to convey cumulatively the flavor and spirit of expanded career roles and the paths that six experienced professionals have taken to get there.

Case History 1: From Technical Writer to Associate Partner

In her career at Accenture (formerly Andersen Consulting)/Arthur Andersen & Company, Janet Borggren has risen from technical writer to associate partner for technology. Her current focus is e-commerce solutions. Her projects have ranged from designing business components for various clients to designing and running back-office operations for a dot-com startup. Because most work is performed at the client site, she continues to travel extensively.

Getting Started. Janet taught high school math for 3 years before joining Arthur Andersen in 1984. She had decided to leave teaching and, after researching new careers, chose technical communication as a field in which she could continue to use her experience in organizing information.

As a technical writer, Janet designed and wrote user manuals, help text, training materials, presentation materials, and marketing collateral for PC-based audit applications. She became documentation manager for an application software organization within Andersen 5 years later. In this position, she led a small team in maintaining more than 60 manuals plus help text for a software package released every 6 months. She also directed initiatives to improve the quality and efficiency of the publishing effort.

Making the Transition.   In 1993, Janet was ready for a new challenge, in part because "lots of tech writing is explaining what isn't intuitive." Andersen was forming a new group to conduct applied research into better ways to build systems and thus improve human performance. Janet became a charter member, spurred by a growing interest in usability evaluation.

The new group, which included interaction designers and integrated electronic performance support system (EPSS) designers, concentrated on user interface (UI) design and usability. It investigated different methods for building applications based on different organizational models and technologies. Because the group also needed to disseminate the new knowledge it developed, Janet designed and conducted workshops on task analysis, object-oriented design, UI design, and usability testing.

Along the way, Janet became an expert in designing the overall architecture and software components for enterprise applications, and she decided to pursue this area. A central skill in her job is the ability to take oral or written information, identify key points, and present them in a structure that reinforces meaning. Also important is the ability to layer information—for example, create 12-second, 12-minute, and 2-hour versions—to help manage complexity. Finally, understanding what technology can and cannot do, and what it can do easily and what it can do only with difficulty, is also essential.

Looking Ahead.   Andersen actively supports career evolution among its employees. (Indeed, the company has reinvented itself as technology, customers, and expectations have all changed.) The firm assumes people want to learn and invests heavily in training. The ongoing assembling and evolving of project teams creates an abundance of opportunities, allowing employees to seek new roles. Strong networking gives employees the courage to take risks because they always know someone they can call for help. Finally, senior people who have insight into options throughout the firm conduct employees' annual performance reviews.

Janet expects her work in e-commerce to remain exciting for a few more years. She laughs as she wonders how old she can be and "sleep in a hotel four nights a week."

Case History 2: From Technical Editor to Usability Labs Manager

Since 1999, J. O. (Joe) Bugental has been the usability labs manager for Sun Microsystems in Menlo Park, California. He oversees a staff of four who run a three-suite usability lab, as well as numerous contractors and firms that provide usability services to the company's development project teams.

Getting Started.   Joe started his communication career as a marketing/public relations writer-editor in the late 1970s, working first in a corporate setting and then freelancing. In the early 1980s, he started learning WordStar while on a contract assignment and soon bought his own microcomputer. His struggle with the documentation led him to perform what today we would call a heuristic evaluation, and he began talking to people at MicroPro about documentation and usability issues. MicroPro responded by offering Joe a position as a technical editor.

From 1982 to 1995, Joe worked in technical publications departments at various computer technology companies. He progressed from technical editor—rewriting the work of seven technical writers for one memorable manual (the WordStar 3.3 User's Guide)—to technical publications manager with 11 people on staff. Although he was drawn to the field by the promise of a steady paycheck and benefits, Joe discovered he had a fascination with computers and a drive to make technology products more accessible to real-world consumers.

Making the Transition.   Early on, Joe had an effective manager named Donna Crawford who was also a visionary. She motivated his initial interest on showing developers how to improve products through usability testing. Not only did she instruct her documentation staff to find real users to inform their work, but she also allocated a budget for usability activities and arranged "master classes" and guest speakers on user interface design and usability.

Joe and his coworkers "assigned" themselves usability testing as part of their jobs. They began informal usability testing in 1983, conducting simultaneous user sessions and then bringing the users together in a focus group to debrief on their experiences. Joe also looked for opportunities to perform informal (and sometimes iterative) usability testing of software and documentation at the other companies for which he worked, where he aligned himself with tech support and quality assurance staff.

Although his last technical publications department wanted to start a usability program, it had no support to do so from the rest of the company. "Usability engineer" was not a defined position. To complete his transition from documentation to usability work, Joe joined Tec-Ed, a usability research firm, in 1996.

As a Sun usability labs manager, Joe finds particularly valuable his theater and personnel background and communication skills, and his understanding of what is realistic within product development and release schedules (he studied software and systems in university extension after completing his degree). His earlier career experience in a computer simulation lab and as an interviewer also helps.

Looking Ahead.   Joe accepts the limits to what he can accomplish as usability labs manager: 100 usability practitioners in a company of 40,000 employees cannot support all of the work being done. Recognizing the connection between insights from usability evaluation and good interface design, he hopes to facilitate a process that will lead to good design earlier in a product's or Web site's development.

Case History 3: From Technical Writer to Usability and Interface Design Manager

Jacqui Miller is the manager of usability and interface design and the lead usability specialist at Deloitte & Touche LLP. Her responsibilities include providing usability evaluation to projects as well as promoting the benefits of usability/UI design services to the organization and managing a group of at least four professionals.

Getting Started.   Jacqui got her first experience in technical communication through the cooperative education program at Drexel University. In addition to coursework on all types of writing and broadcast media, she worked as a computer technical writer at the Wharton School of Business and for a small communication company.

After graduating in 1987, Jacqui took a position with a large company that develops financial and other software for institutional food programs. For 2 years she was the technical writing staff, serving as writer, editor, graphic designer, trainer, software QA, and user support—roles she could perform only because of her co-op experience. She started to "bunk out" with the developers on the 15th floor to bridge the gap between them and her department on the 31st floor. She also became a member of the Society for Technical Communication (STC) and used it as her support group.

Jacqui joined Deloitte & Touche in 1989 as senior technical writer in the Princeton office with responsibility for audit software documentation. During her tenure, she redesigned and rewrote the 750-page user manual, introduced online help to the product, and offered editing services to senior managers and auditors.

Making the Transition.   In 1994, seeking new challenges and more product variety, Jacqui moved to Deloitte's central technology group in Hermitage, Tennessee. She started as technical editor on a team of technical writers and designers brought together to produce multimedia, but soon thereafter became team leader for up to 18 regular and contract staff.

In 1995, Jacqui approached her manager about offering usability services. The manager agreed to support her idea if Jacqui could sell the service within Deloitte. Over the next year she read extensively about usability, did internal marketing and education, started training some writers in usability techniques, and picked a few clients most likely to benefit. The pilot projects were a success, and Jacqui began doing usability work full time.

Jacqui is somewhat ambivalent about management—helping staff members develop is rewarding, but hiring and firing are difficult. Her greatest joy is collaborating with her group in hands-on projects. Attention to detail and the ability to chunk information, write clearly, and improve design on both the printed page and computer screen are the skills she uses most. Plus, she is not afraid to speak up when decision makers do not have all the information they need to make good decisions.

Looking Ahead. Jacqui is still learning about how best to combine usability and UI design services, as well as how best to sell these synergistic practices to her internal clients. She is intrigued by the idea of a chief user experience officer (CXO) having final authority for the way a company's interfaces are presented. But she thinks it may take a "Ken Blanchard of usability" to motivate and excite business owners, chief information officers, and vice presidents before this position takes hold.

Case History 4: From Junior Technical Editor to Marketing and Web Content Writer

Sally Hanna works as a marketing and Web content writer at a Web consulting company that designs and builds Web sites, intranets, and Web-based e-business and e-commerce applications for other companies. Despite her title, she spends more time architecting than writing as she creates visual outlines of large Web sites and storyboards the pages.

Getting Started. Sally got her start in technical communication in 1977 with a large architect/engineering firm that builds power plants. Hired as a secretary, Sally soon was editing and improving her department's documents. Her manager—who was also the technical publications manager—decided to give Sally a chance as a junior technical editor. A promotion to technical editor quickly followed.

Sally worked first on documentation for nuclear power plants, an experience that taught her she could edit text she did not fully understand. She then moved into records management for a fossil-fueled power plant, which exposed her to computers and databases—and a lot of poor documentation. To improve work quality, she wrote procedures for the clerks who processed the tens of thousands of documents and drawings. She also continued to edit for the engineers who sought her out.

With the downturn in power plant construction, Sally approached a computing company in 1985. A contact she had met in a technical writing class at Eastern Michigan University hired her as a senior technical writer. Assigned to document early PC-based tools for business, Sally found her understanding of organizations to be as important as her writing skills. She was able to write primarily for end users, because this audience needed the most assistance with understanding software.

Making the Transition. Sally joined Tec-Ed in 1987 in search of career growth, which she felt she could get with this company because "the book was the product." (Tec-Ed at the time specialized in computer documentation and training.) She wrote reference manuals, user manuals, and training materials and had her first taste of marketing writing.

From 1989 to 2000, Sally worked freelance "to own my life," expanding her services to include online help and enjoying the variety of projects and clients. In the late 1990s, she began to experience a rising frustration because she was increasingly contracted to produce documentation for programmers and system administrators rather than the nontechnical users with whom she had more sympathy.

As she considered how to reconnect with users, Sally found a Web listing for a marketing and technical writer. The position reported to a woman she knew from the dance community to which she belonged and whose newsletter she had edited. That woman is now her manager and mentor.

Variety characterizes Sally's current position. In one early project, she participated in a business analysis of a client's needs, translated her findings into functional specifications for the user interface of the client's intranet, wrote the user guide after programming was complete, and performed quality assurance on deliverables. She usually works on several proposals a month ("It's still information design"), thus getting the jump on new projects. She has also begun to learn HTML, become a "process preacher" in this young company with many first-job employees, and found satisfaction in mentoring others.

Looking Ahead. Sally's first year in the rapidly growing company has cycled frequently from stimulating to frightening, and from frustrating to satisfying. She calls the transition to site architect "very tool-based," which she terms "terrifying" because of her drive to perform at a professional level from the start. Although the concept of structuring information is the same as in writing, she notes, "The medium is more visual and the tools were new to me." On the other hand, she relishes the chance to learn by doing and is willing to take intellectual and emotional risks.

Case History 5: From Senior Technical Writer to Business Operations Strategist

Denise D. Pieratti has spent most of her technical communication career at a multinational corporation that specializes in document solutions. Most recently, she developed strategies for customer business operations, investigating and recommending ways to grow the business and increase revenues through effective management of product-use information.

Denise has moved to Nortel Networks, where she leads part of a team developing applications for wireless Internet devices. She is responsible for the content delivered by traffic, weather, directions, advertising, and other applications to cell phones, PDAs, and laptop computers, and for influencing call flow/interaction and the user experience.

Getting Started.   Before becoming a technical communicator, Denise worked as an engineering geologist on major facilities such as power plants, dams, and tunnels. She rose to project geologist (project manager) before hitting the glass ceiling; she was laid off during a downturn in the early 1980s. As she looked for a new career, Denise met someone from STC who told her about technical writing. Further research convinced her she could combine her love for computers, communication, and creativity in this field. She took her first position in 1984 as senior technical writer with the document solutions company.

Within 6 months Denise was managing a three-person team that was developing software manuals for a new production publishing system. Six months later she moved from Los Angeles to San Diego to establish a technical publications department for a line of business that included eight products and 40 manuals.

Making the Transition.   Denise's department was one of the first in the company to produce online help. In addition, because it matched the target audience for one of the products, the group started performing usability evaluation after Denise attended Judy Ramey's 3-day usability testing workshop in 1987.

In 1990 Denise became a training program evaluator charged with assessing technical training for people selling production publishing products. The job was highly structured and tightly focused, so to inject "more fun and exciting things" into her life, Denise entered the master's program in technical communication at the University of Washington. Discovering rhetoric changed her life, as she became aware of the process and power of writing. Also, learning more about usability opened new career opportunities.

Denise moved to Rochester in 1994 to manage a new documentation/usability group for internal business applications. When the group was disbanded, she became a marketing communication manager. She was awarded a social service leave 2 years later, returning to the company as a business operations strategist.

Looking Ahead.   Denise credits her ongoing success to several factors: She communicates technical information easily and effectively, bridging the gap between technical experts and nontechnical audiences. She cares passionately about the impact of technology on people, not only on how people use technology but also on how it affects their lives. Thanks to her training in geology, she knows how to take information from different places, make connections and links, and look for holes and patterns. She does not hesitate to make hypotheses when she does not know all the answers, and she is fearlessly inquisitive.

Case History 6: From Policy Development to Product Marketing and Management

Helene Schultz is responsible for the management and marketing of several strategic software products at a company that applies information technology to e-business solutions. She makes decisions about the products' R&D funding, packaging and pricing, and marketing collateral, and makes sure development, documentation, manufacturing, and sales are in place for product release.

Getting Started.   In 1985 Helene accepted a position as policy development manager with a hospital. She thought she was getting into health care administration, but instead found herself the sole technical communicator documenting administrative policies and procedures of a three-facility hospital and long-term care unit. The policies and procedures were required for business reasons as well as for accreditation. The three facilities had merged into one large hospital corporation, which needed to consolidate services and maintain quality standards. The hospital's senior management recognized documentation would help them succeed with the merger.

Helene's next job, at a student loan servicing company, introduced her to computer documentation. She prepared policies and procedures for student loan programs, user guides for mainframe and PC applications, and technical training material.

Helene joined her current employer 5 years later as a contract writer. She became a regular employee after residential moves out of state and back, during which time she worked in technical communication for manufacturing and retail applications. She created user guides, online help, and marketing materials before being promoted to the product information management team in 1997.

Making the Transition.   As a member of this management team, Helene carried out department- and company-wide directives and had some influence over how the department was run. She tried to advocate on behalf of the user community, which was difficult because she knew little about the company's customers. She became interested in marketing as a way to learn about users' requirements, which she could then share with product developers and product information specialists.

Helene transferred to her company's marketing division early in 2000. With no previous experience in product management and marketing, she sees herself as still learning. She plans to supplement a 2-day class on software product management and marketing from an external vendor with marketing management courses at a local college. Although she reads extensively, she is skeptical she can learn all she needs to know from books when the software product market is so competitive. She is prepared to try new approaches based on her previous experience and intuition.

Helene's master's degree in political science, with an emphasis on public policy analysis and administration, provided good training in locating and understanding large amounts of information. She finds that good communication and organizational skills are as important in her current position as in any technical writing position. In fact, because she writes her own marketing communication, she considers herself still involved in technical writing.

Looking Ahead. Helene characterizes the transition to a new role as "always painful." She feels simultaneously optimistic and overwhelmed by all she still has to learn, and looks forward to continued career development in product management, project management, and personnel management.

SUPPORT FOR TRANSITIONS

Our interviewees offered advice for professionals working in technical communication today and for academic programs educating students who will enter the field. In some cases, the advice about preparing for a transition to a related field applied to both groups.

For example, Joe Bugental observed that with the emphasis on Web applications, "product" and "documentation" have merged—everything is "information" now. Technical communicators have opportunities to integrate instructions with application content, and their responsibility for both is becoming acknowledged.

Our interviewees agreed that good communication skills, the ability to respond rapidly to change, and multitasking skills are vital in business. People in technical-communication–related positions must be able to talk with everyone from developers to chief financial officers, and to target messages to specific audiences. They must know enough about technology to understand what it can and cannot do, and to understand the technical limitations that developers face. Finally, they must appreciate business needs as well as user needs on which technology decisions are based.

Above all, both novices and veterans need to know themselves, understand what energizes them, and be alert to how the world is changing because—as Janet Borggren points out—the next thing may not have been invented yet. That has certainly been her experience: None of what she does professionally now existed when she began her career in technical communication.

Advice for Working Professionals

Our interviewees also offered the following advice for working professionals.

Develop a Professional Specialty or Domain Knowledge. Technical communicators need more than writing and interviewing/research skills, regardless of the transition they are contemplating. They also need a professional specialty (such as information architecture, usability evaluation, user interface design, Java programming, or project management) or domain knowledge (such as biotech, finance, or wireless devices).

Take Advantage of Education and Training Opportunities.   Technical communicators should read books and journals, enroll in courses, attend conferences, and ask questions of experts in their related field of interest. It is important to look at a range of products and Web sites to develop ideas, and to keep current with new technology and tools.

Get Hands-On Experience.   Technical communicators can approach their managers about getting on-the-job training in a related field. Companies are often willing to take a risk on employees who show initiative and self-discipline and who learn quickly.

For example, technical communicators who want to become project managers can learn by running some kind of project. Although different project management methodologies exist, knowing a particular methodology is less important than understanding the dynamics of the process. Technical communicators interested in product marketing should serve an apprenticeship to develop not only techniques and research skills, but also familiarity with and commitment to the product or service being marketed.

One interviewee suggested that technical communication departments model themselves after technical support departments. That is, technical communication departments should expect that they are preparing staff for other roles in the company as they hire bright people and expose them to new technology in the course of mastering the company's products.

Get to Know People Throughout the Organization.   Technical communicators must avoid an us-versus-them mentality and instead form relationships with people in other departments. For example, getting to know the developers enables technical communicators to share expertise, do their jobs more effectively, and increase the chances of influencing product development. By making contacts throughout an organization, technical communicators can identify potential career-building opportunities and avoid being labeled as "just a technical writer."

Identify Mentors.   Technical communicators should identify mentors and develop a network of people who know about them, care about them, and want to look out for them—in short, people who can answer questions, offer insights, and help them find the next job.

Understand the Company Culture.   An understanding of the company culture must complement good communication and people skills. Some organizations value a lively exchange of ideas; others perceive disagreement as negative and label dissenters as "not team players." Disrupting the environment could disrupt one's career progress.

Make Professional Contacts Outside the Organization. Technical communicators need to join professional societies in their areas of interest and be active in organizations. Especially useful is to volunteer for leadership positions. Another approach is to meet informally but regularly with people already working in other fields, as a way to ask questions and hear war stories.

Change Jobs Every Few Years. To have a rich and varied career path, technical communicators must be prepared to change jobs every few years. If a job has become too comfortable or is not fun any more, it may be time for a change. When that time comes, technical communicators need to be clear about what their skills are in order to sell those skills to others. To build credibility, it is also important to be frank about the extent of one's expertise in a new field.

Advice for Technical Communication Programs

Our interviewees offered the following suggestions for improving technical communication programs.

Prepare Students to Think at an Abstract Level. Many of our interviewees have been well served by a liberal arts background. With that in mind, Janet Borggren cautioned technical communication programs not to become like trade schools and focus on techniques. Instead, they need to prepare students to think abstractly. It is unclear how long any particular fact or skill learned today will be useful; innovation requires thinking beyond techniques. Janet also recommended that students take courses that are neither technical nor communication for "freshness and cross-fertilization."

Provide Firm Grounding in Fundamentals and Exposure to Different Fields. It is impossible for technical communication programs to educate students completely with the world changing so fast. By providing firm grounding in the aspects comprising technical communication today, plus exposure to related specialties and business principles, programs can produce well-rounded graduates. Programs might even consider requiring students to study in some depth a professional specialty or subject domain of their choice.

Denise D. Pieratti, a graduate of the University of Washington's master's program in technical communication, strongly recommends that students take at least one class in rhetoric. Rhetoric taught her that "how you present something—how you write it, how you package it, how you communicate it—can make people change their minds and take action. That's powerful stuff and why I take all my writing, even a half-page memo, very seriously." She also recommends taking classes in cognitive psychology, organizational dynamics, user and task analysis, usability, and e-commerce (or the currently hot technology topic).

Coach Students in Interpersonal and Behavioral Skills.   Because their success in business will depend partly on whom they know, students need to learn how to approach people who have power, authority, and information; how to navigate the chain of command; and how to negotiate. They also need to learn how to work in teams, how to establish and maintain good internal relationships, and how to find out about users. Finally, they need to learn how to manage their time and how to manage their careers to avoid exploitation or burnout.

Offer Internships to Prepare Students for the Business World.   Technical communication programs should prepare students to operate within an entire business organization, not just a technical communication department. The best way to prepare students for life in an organization is through cooperative programs and internships.

Support Working Professionals as Well as Full-Time Students.   Depending on their age and prior experience, students will have different motivations for enrolling in a technical communication program—from keeping up with changes in the field to job training. With the importance of life-long learning in today's economy, technical communication programs should offer courses that help working professionals upgrade their technical skills and knowledge.

CONCLUSION

What do we think these case histories and suggestions for career transitions tell us? Our interviewees wanted to make a difference, and they succeeded by being restless, inquiring, relentlessly curious, and intellectual. They have taken initiative and they have taken risks. In addition, they have cultivated professional relationships throughout—and often outside—their organizations.

Based on this small number of interviews, our personal experiences, and those of colleagues (Humburg, Rosenbaum, & Ramey, 1996; Rosenbaum, Rohn, & Humburg, 2000), we see some patterns:

- Interest in user experience expressed by at least one high-level manager contributes greatly to our ability to have strategic impact. Even midlevel mentoring helps.
- Telling stories is vital; one reason that technical communicators have the potential to gain strategic influence is that we can be missionaries and communicate the benefits of our desired goals.
- In U.S. organizations, technical communicators need not wait to be invited; we should suggest new roles and promote change, rather than wait for management to direct it.

What should our field conclude about the career transitions open to technical communicators? Even though we believe it is too early for conclusions—we have not collected enough data—we do recommend that technical communicators honor, embrace, and study members of our field who have moved to influential positions within their organizations. Invite them to speak at meetings, learn what characteristics of their organizations helped or hindered their career growth, ask them to mentor others.

When studying these technical communicators, we should uncover more systemic reasons for their success, and learn what characteristics of their organizations contributed to or hindered their progress. We consider such research important enough to merit support by research grants. We also strongly encourage in-depth qualitative research on how technical communicators can be more influential and effective in organizations.

To conclude, people with degrees in architecture or dentistry rarely practice as anything but architects or dentists. By contrast, technical communication can support transitions to influential positions in organizations because technical communicators have the potential to master a succession of different specialties, continually evolving their expertise. Each specialty becomes embedded in a professional's experience, and the emerging relationships strengthen support networks in the field of technical communication.

Currently, the technical communication profession does not formally recognize and mentor such transitions as a natural progression. Instead, sadly, we often view these people as lost to our profession, rather than as our compatriots. By redefining the boundaries and roles of technical communicators, we will ensure the growth and influence that our field deserves.

ACKNOWLEDGMENTS

The authors gratefully acknowledge the professionals who shared the stories of their careers for this chapter: Susan Bonnette, Janet Borggren, J. O. (Joe) Bugental, Roslyn Fry, Sally Hanna, Jacqui Miller, Denise D. Pieratti, Mickey Reilly, and Helene Schultz. We regret that we could not include everyone's case history or more details in those that do appear.

REFERENCES

Humburg, J., Rosenbaum, S., & Ramey, J. (1996, April). Corporate strategy and usability research: A new partnership. *CHI 96 Conference Companion,* 428.

Rosenbaum, S., Rohn, J. A., & Humburg, J. (2000, April). A toolkit for strategic usability: Results from workshops, panels, and surveys. *CHI 2000 Conference Proceedings,* 337–344.

10

Advancing a Vision of Usability

BARBARA MIREL

University of Michigan
Mirel Consulting

A nurse on an acute care hospital floor wheels her medication cart into a patient's room, careful not to jostle the laptop and radio frequency scanner that sit on top of the cart. She parks her cart, taps the icon on the touch screen of the computer that opens the bar code medication program, and logs into the program. She then scans the patient's identification number and, in response, the bar code medication program displays the patient's record—a list of his prescribed drugs for this scheduled medication pass. One by one, the nurse takes the patient's drugs from a cart drawer and scans the bar code of each. As soon as a bar code matches a prescription for this patient ID, the system "bings," marks the drug approved on the screen, and documents the match, presuming it to be an accurate and successful drug administration. All eight medications for this patient "bing" positively. The nurse fills a cup with water, gives it to the patient along with his pills, and closes his record. The process took moments; the right patient got the right drugs at the right time and place; and the nurse moves on.

In this situation, the bar code medication software is highly efficient, effective, and usable. The program readily opens to the patient's record; the screen provides quick and easy access to the drug names, times, dosages, and approval markings. The user signs on, then opens and closes the record simply. The functionality for administering and documenting all and only the right drugs seems comprehensive; and, true to its objective, the program guards against human error and assures patient safety.

This usability picture grows dim, however, once patient cases are not textbook perfect. When the slightest deviation occurs, the software frustrates rather than supports nurses in their work. One such case involves a patient who cannot swallow easily so that the nurse needs to crush the pills and mix them with applesauce. This patient has 13 pills scheduled for the morning medication pass. All goes well at first. When scanned, all 13 drugs "bing" positively. The nurse begins feeding the patient the applesauce mixture but the patient is only able to swallow a quarter of it. Now usability takes a nosedive. The nurse has to edit the record documenting the drug administration to undo the automatic entry for successfully giving medications that the program entered for each positive scan. But to edit, the nurse must access the documentation from a centralized patient record system that runs on a different operating system and that interfaces with a different program. The nurse spends 5 minutes logging into the other program, hung up by a painfully slow and underfunded hospital network. Once in the other program, she has to edit each of the 13 medications separately, another time-consuming process. It is exceptionally long because each entry requires overriding the predefined editing options to note that in this unusual case, the patient took some but not all of each pill.

Other cases introduce different complications, each requiring its own set of interactions. One patient, for example, has a tapered dosage, meaning that the nurse must clinically judge the amount to give based on 72-hour-long patterns that she has seen in the patient's pain levels, vital signs, and cumulative prior dosages. The nurse has to gather this information from at least three sources, one within the bar code medication program and two outside of it. Retrieving the necessary information requires many navigation steps and query procedures, but it is only one part of what quickly becomes an excessively difficult task. Most taxing are the actions and memory load required for copying relevant data into one display, arranging it for easy interpretation, and deriving the totals needed for a diagnosis.

These "exceptional cases" are more common in everyday nursing than are the textbook perfect cases. Yet the bar code medication program does not accommodate them readily. The problem is not omission. The bar code medication program has the functionality, features, and interface controls to enable users ultimately to get the information and commands that they need. But nurses are not willing to go through the extensive processes that the program and other related information systems demand. They object to the many scattered and piecemeal actions required for putting together what to them is a single "chunked" task—for example, analyzing vital sign history. They object to the inordinate amount of time spent working the tool and remembering data. It compromises the efficiency on which they are judged in their performance evaluations, and it takes away from the time that they can spend on the personalized bedside care that motivated them to go into nursing in the first place. Faced with a program at odds with their professional culture, organizational practices, and technology infrastructure, many nurses devise workarounds or cut corners that unintentionally can introduce new risks to patient safety. As a result, poor usability due to a mismatch with actual work practices and needs can truly be a life-and-death matter.

This case with the bar code medication program is not atypical of software design for complex problem solving. Unfortunately, it is not some isolated incident of design gone amiss. What happens in software production contexts to cause this large disconnect between usability in letter-perfect situations and in situations that are messy, conditional, or idiosyncratic? Questions such as this are gaining urgency as software companies increasingly become customer- rather than technology-driven and realize that software must fit the messy realities of users' complex work-in-context. Software development teams often are unsure about what it means to build for this kind of usability or what it should look like. As a result, usability specialists rapidly are being brought onto software projects. However, positive as this trend is, it is not enough to assure a lead role for usability in enhancing the success of a product. To assume and exert leadership, usability specialists need to be agents of change, a role they have not yet assumed in most workplace contexts.

As technical communicators increasingly move into usability roles—in the software industry and elsewhere—they could readily assume this leadership. They are trained, perhaps as are no other specialists in human–computer interaction, in the rhetorical perspectives necessary for effectively matching the media and design of software support to particular audiences, purposes, activities, and contexts. Yet, in many companies usability leadership is sorely lacking.

In this chapter, I argue that as leaders, usability specialists have to introduce a new vision of what it takes to support complex work-in-context, discussed in more detail later as *designing for usefulness.* In particular, if usefulness is to take center stage, a shift is needed in analyzing and designing for complex tasks. This shift involves moving from task- and even user-centered designing to designs centered on use-in-context. Making this shift depends on strong usability leadership because it requires new ways of thinking and doing. Usability leaders have to bring about innovation and change in task analysis, task representations, and development processes.

This chapter examines the vision, approaches, and changes involved in designing for usefulness in one type of environment—contexts that produce software for complex tasks. The bar code medication software mentioned earlier is one such program. Others include programs that support, for example, product planning, analysis of profitability, or allocation of resources. In these environments and for this end of supporting users' complex work, the chapter examines issues relevant to bringing usability leadership to bear on the design and development of the software.

WHAT IS USABILITY?

"We're user-centered but we're pretty vanilla about it—you know, the usual interface improvements—because of constraints. We figure it's still better than it might have been."   —Usability Director, Business-to-Business web development firm

Fifteen years ago, Gould and Lewis (1985, p. 300) defined usability, arguing that "any system designed for people to use should be easy to learn (and remember), useful, . . . contain functions people really need in their work, and be easy and pleasant to use." Usability specialists since have elaborated on the qualities identified in this definition, but the scope of what usable programs should do has not changed from this succinct description. Usability, as Gould and Lewis note, involves multiple dimensions—ease of use, ease of learning, pleasantness, and usefulness; all combined, they provide users with a positive work experience. Each dimension is equally necessary for assuring that users seamlessly integrate a program into their ongoing work. All are intricately intertwined. None is independent of the others.

Yet, in many development contexts, a comprehensive vision of interrelated usability dimensions gets broken apart. Each dimension—ease and efficiency of use, learnability, enjoyment, and usefulness—becomes a separate objective. The comprehensive whole becomes a pick list of options. Design and development teams choose to build for some dimensions while neglecting others based on project deadlines, resources, and other constraints (Grudin, 1991). From a pick list perspective, team members are likely to accept without question a rationale for designing the bar code medication program that might go something like the following: The program should read in and display patients' vital signs from other programs. But we do not have the architecture or headcount to build that cross-program integration into this release and ship on time. So we will be user-centered in the interface and provide menu options and buttons that will let users easily access the path that they have to take to get out of our program and into another to get the data that they need.

Despite such expressed user-centered intentions, actual users in context are hardly grateful for such make-do ease of access, and they are hard pressed to find these programs user-centered. Rather, they get exceedingly frustrated with the program. Put simply, it does not do what users want for the work that they have to do.

In these and other similar cases, design choices and rationales are out of sync with users' holistic experiences of usefulness. This phenomenon is one of the core issues that usability leaders must raise and counter. In the midst of doing their work, users experience usability as a component of the total program quality. Designers, by contrast, often treat usability as a set of discrete parts, ready to be traded off without resistance as soon as the "too costly" behemoth rears its head. Cowed by budget or personnel constraints, development teams routinely forgo hard but useful solutions without debate and opt for "vanilla" ease-of-use interface designs such as navigation buttons. They rationalize that by offering improved ease of access at the interface, they are delivering user-centered programs.

Usability leaders must distinguish between ease of use and usefulness. Ease of use involves being able to work the program efficiently and easily; usefulness involves being able to use the program to do one's work-in-context effectively and meaningfully. Stripping usefulness from ease of use and focusing primarily on the latter is an incomplete recipe for usability or user-centeredness.

Tradeoffs are inevitable in design and development, but they require critical assessment and debate (Rosson & Carroll, 1995). They require posing and answering such questions as: Is a sacrificed or neglected capability negotiable from a user's point of view or is it critical to integrated work practices? Is it better in the long run to invest more in development now and release later, than to provide only make-do solutions in order to release now?

To adequately position usability in these debates, strong leaders are needed. They need to overcome teammates' piecemeal notions of usability and show that partial usability is no more favorable to users than partial system performance. To do so, usability leaders need to bring in empirical data on users' needs, practices, and boundaries of tolerance. To build a convincing case from these data, however, leaders first need to lay a groundwork. They need to clearly show how the dimensions of usability are related to each other for a given product and what this relationship implies for software design for complex tasks.

THE PRIMACY OF USEFULNESS FOR COMPLEX TASKS

"Requirements are discovered through the contingencies of everyday use."
—Suchman, 1997, p. 56

"A designer who is thinking, 'how do I decrease the number of keystrokes,' often ends up finding and improving lots of little problems with all these improvements adding up to little. By contrast, a designer who tries to craft an interface that elegantly fits a user doing the task often ends up sidestepping many more problems in one fell swoop by neatly eliminating the task actions in which those problems lived."
—Dayton, quoted in Strong, 1994, p. 17

The ways in which the dimensions of ease of use and usefulness relate to each other vary by product and by the work, users, contexts, and purposes that are supported by the product. In software for complex work, this relationship needs to give primacy to usefulness. Complex tasks characteristically vary with context; the same task is rarely if ever performed the same way twice. In complex tasks the means for performance are not entirely known at the outset. They emerge and become more specified as people explore conditional factors relevant to the task. Knowing what to do next requires coordinating, arranging, and relating relevant factors. Typically, several renditions of acceptable arrangements exist. Similarly, the rules, heuristic strategies, or trials-and-error that people apply lead to processes that allow for several outcomes. Complex tasks, therefore, cannot be formalized into fixed procedures and rules. Programs cannot presuppose rules and map formulaic steps onto program features and commands. Rather, software for complex tasks has to be flexible enough to "lean on" the knowledge and frames of reference that users bring to it and to adapt to a range of situational factors, emergent conditions, goals, and moves (Agre, 1997). Designing for this flexibility and variation in work-in-context is designing for usefulness. For complex tasks, usability leaders need to work with program designers and developers to understand and figure out how to support the range of users' possible approaches in context. Correspondingly, they need to understand the ways in which various configurations of contextual factors and relationships shape the range of actions that are possible and the choices that people make in a given instance. Designing for usefulness assures that the right sets and structures of interactivity will be in place for the right users to perform the right possible actions for the right situations. When software is effectively designed for usefulness, it provides and displays a framework for task performance that is consonant with the goals and situational "markers" that trigger users to conduct their work in specific ways. After this framework is established, making interactivity easy, quick, accessible, understandable, enjoyable, and navigable falls into place. To do otherwise—that is, to address ease of use first without initially assuring usefulness—increases the risk of making the wrong model of a workspace easy to use and, therefore, of little pragmatic value.

Unfortunately, designing for usefulness is usually given short shrift in production contexts. Design teams are prone to jumping prematurely to detailed specifications and ease-of-use concerns. One reason may be competitive pressure. Teams hurry to establish features and operations so that they can get on with coding and usability testing and ship before competitors do. Even if competition were not a factor, however, foregrounding usefulness is more the exception than the rule because it runs counter to several solidly rooted, conventional design practices. These include teams' tendencies (a) to analyze and design for tasks at a low level of detail; (b) to assume a procedural or operational orientation to representing and supporting user tasks; and (c) to exclude usability from the critical junctures in the development cycle in which decisions are made that either leave open or close later design possibilities for usefulness.

Each of these three tendencies is inappropriate and counterproductive for building the support that users need for complex tasks. Each needs to be addressed and redressed by usability leaders. I now turn to new approaches and challenges associated with each.

DESIGNING FOR USEFULNESS:
APPROACHES TO TASK ANALYSIS

"There is a tremendous difference between designing for function and designing for humans."   —Cooper, 1999, p. 89

"We're all smart people here. We don't need to go out and do all these studies. We know what to build."   —Chief technology officer at a project team meeting

As noted earlier, in software contexts at present, a common approach to analyzing and designing for users' tasks is to break them down to their smallest parts and rules. At this unit level, tasks are the simplest actions that users can handle without going into problem-solving mode and that require no control structure to accomplish (Green, Schiele, & Payne, 1988). By decomposing performance and its rules and resources, analysts can represent users' activity as a hierarchical flow of goal-driven actions and knowledge down to the smallest set of tasks. At this low level of analysis, they can design, specify, and implement corresponding program features and objects. These programs represent users' work by representing its composite parts. The underlying assumption is that the whole of users' activity equals the sum of its parts.

Increasingly, many design teams are resisting a formal decomposition method of task analysis and design (Bever & Holtzblatt, 1998; Kies, Williges, & Rosson, 1998; Rudisill, Lewis, Polson, & McKay, 1996; Wixon & Ramey, 1996). Bever and Holtzblatt (1998), for example, suggest five frameworks for representing contextual inquiry findings that capture relationships among work processes, social roles, environmental dynamics, and physical arrangements. These researchers' representations take the form of workflows, task sequence and goal diagrams, models of artifact use, cultural dynamics models, and physical layouts of the work environment. Yet these and other researchers' representations often end up unintentionally designing for unit tasks. These teams pursue alternate contextual and ethnographic methodologies and represent tasks as interrelated user cases, user profiles, and models of workflows, task sequences, and communications and culture. However, when these teams move from contextual descriptions to object-oriented design and programming, such factors as situational influences, contingencies, and high level conditional interdependencies become too hard to capture in design. Teams end up focusing instead primarily on separate elements of the work, ultimately mapping them in one-to-one fashion to program features and controls. Contextually oriented designers struggle with finding methods to use to move from task descriptions to design without ending up with a focus on discrete, context-free, low level actions (Wood, 1998). As detailed in the next section, it is unlikely that any methodology will achieve this end unless designers first change the framework of the task representations that they compose.

In regard to level of detail in task analysis, usability leaders need to continue to promote a contextual vision and ensure that this vision does not devolve into designing for dissociated low level operations. To do so, they first must reveal why the dynamics of complex tasks necessitate against summing parts to a whole. They then need to compose task representations that do not readily lend themselves to being dissected into elemental actions. This section addresses why summing unit tasks is inappropriate for complex tasks. The next section examines ways to compose task representations.

Regardless of how a unit-level focus occurs, complex tasks can neither be supported nor conducted as a sum of unit parts. Usability leaders need to argue that complex tasks are complex because they are not comprised of well-defined, rule-based, serial steps that cumulatively and predictably sum to the whole of a task. Rather, they are emergent and dynamic. As noted earlier, many components interact with and adapt to one another—cognitive, behavioral, situational, and technological components—creating problem spaces in which a number of moves are possible. Interdependencies across components and unexpected side effects from their interactions make it impossible to sum parts to a whole. At any given point, people choose some moves and not others. These choices reconfigure arrangements between components, setting into play new constraints, interactions, and effects. In computer-supported complex tasks, much of the difficult processing needed for resolving a task or problem has to take place on the screen and be controlled by users rather than going on invisibly "beneath the hood." Users have to process and interact with displayed information to gain the insights needed for transferring knowledge to similar tasks and to shape analysis as they like. Or, another way to view it is that, in complex tasks, users need to learn to fish, not simply be fed the fish.

For example, a marketing analyst for a coffee manufacturer, analyzing the potential of a new product, needs to view, process, and interact with the data directly. To break into the high-end espresso market with a new product, the analyst examines as many markets, espresso products, and attributes of products as is deemed relevant to the company's goals and analyzes as many as available technical tools and cognitive limits permit. Looking at these products, the analyst moves back and forth between big picture and detailed ("drill-down") views. The analyst assesses how espresso has fared over past and current quarters in different channels of distribution, regions, and markets, and imposes on the data the analyst's own knowledge of seasonal effects and unexpected market conditions. For different brands and products—including variations in product attributes such as size, packaging, and flavor—20 factors or more might be analyzed, including sales revenue, volume, market share, promotions, and customer demographics and segmentation. The analyst arranges and rearranges the data to find trends, correlations, and two- and three-way causal relationships, and then filters data, brings back part of them, and compares different views. Each time, the analyst gets a different perspective on the lay of the land in the "espresso world." Each path, tangent, and backtracking move helps to clarify the problem, the goal, and ultimately the strategic and tactical decisions. In such complex tasks, instead of people attending temporally to a stable territory and clearly mapped action sequences, they spatially survey an unstable territory of complexity, understand its structure, and repeatedly manipulate arrangements and relationships. All the while, they progressively are reducing uncertainty.

Inherently, complex tasks embody uncertainty. Goals for complex tasks are often broad, vague, and revised as inquiries and insights emerge. Problem spaces are not well bounded. Relevant information may be spread across many problem spaces and resources of knowledge at once. Relevant information, moreover, is conditional, multidimensional, and multiscaled. It takes on different meanings depending on the perspectives that people take, and, in complex tasks, people take several perspectives because they have to account for contingencies and conditional relationships. Courses of action are dynamic and emergent, exploratory and opportunistic. At decision points, people face contending legitimate moves, and within the degrees of freedom that they have, they choose based on situational and subjective criteria. As these traits of complex tasks suggest, the structure of complex work is more a dynamic feedback mode than serial behaviors with clearly defined starting, stopping, and decision rules.

Task analyses and designs that attempt to represent complex tasks as linear, decomposed unit actions and rules misrepresent these tasks. In doing so, they often lead to programs that undercut people's abilities to do their work effectively and productively. For complex tasks to succeed, users must have optimal control over information, perspectives, paths, possible actions, and criteria for choosing. Programs diminish this control when they represent complex tasks as a composite of discrete unit tasks without integrating actions and situational arrangements or without calling forth relationships between resources. Such programs overdetermine interactivity, underestimate the scope of people's work, underrepresent important contextual dynamics, and underdevelop the strategies and interactions that people use for getting "from here to there," monitoring progress, and managing emerging knowledge. Usability leaders need to argue against developing programs that model complex work through a built-in prepackaging of means and ends. Ideally, their arguments are based on empirical data from user sites.

Some usability professionals may blame a piecemeal approach to design on the dominance of engineering in a production context and its object-oriented programming. Admittedly, a low-level, unit orientation is needed once product development moves to the stages of detailed specifications and programming. Object-oriented programming does require attributing elemental properties and events to low-level objects, be they things or acts. Yet the constraints of object-oriented programming and design do not force usability specialists or designers down a slippery slope of designing principally for discrete low-level actions and operations. In fact, object-orientation emphasizes giving users support in taking whatever actions and order of actions that they want while maintaining a focus on the primary task (Pancake, 1995). Equally important to properties and events in object technology is the need to carefully construct appropriate relationships within and across classes of objects.

The meaning of complex tasks and task actions lies in relationships. If object-oriented designers and programmers lapse into assuming that whole task meanings are deduced from the sum of properties inherent in component objects, this assumption is not intrinsic to object technology. Object orientation does not foreclose a software team's opportunities to design for situated tasks and integrated actions.

In fact, constraints inherent in object technology are less an obstacle to maintaining an integrated and situated view of complex work during design than is the composition of task representations. Ultimately, object-oriented designing and programming take shape from the task representations that usability specialists bring to the design table and the interpretations drawn from them.

Usability leaders need to assure that these representations are framed around task structure and the structure of functional relationships and interactions. Framed in this way, task representations provide an organizing structure that signals socially shared repertoires of moves and intentions for a given type of problem in a particular time, place, and set of circumstances. Structurally framed task representations are spatial. They emphasize the arrangements and patterns of actions, rules, objects, and interactions that cannot be severed from one another and distilled into discrete parts if users are to see in the program their notions of their work and conduct it seamlessly. Current contextual models do not guard well enough against this severing and distillation.

Promoting and implementing a structural framework for task representations is likely to meet resistance. Usability leaders are apt to find that this orientation challenges assumptions that many people in a production context have about what it means to support users' tasks.

DESIGNING FOR USEFULNESS: APPROACHES TO SUPPORTING AND REPRESENTING TASKS

"As developers, we're freaks. Ordinary people don't think like us. Our users don't think like us. We have to learn how they think."
—Vice President of Development in a software firm
at a Development Group meeting

The preceding sections stressed that conventional models for representing tasks and the software designs derived from them are inadequate for users' complex problem-solving activities. Whether in task models or in resulting interface designs, aiming for a set of prepackaged actions that are tied to precomputed plans does not support users' critical needs for inquiring into open-ended, ill-defined problems. Complex problem solvers need and expect leeway in choosing from available options the best courses of action for their purpose, time, and place, and they need to control the actions that they choose and the information with which they work. To capture and design for the interactive forces in problem solvers' workspace—forces that condition the degrees of freedom and control needed at various points in problem solving—usability leaders need to apply a conceptual framework that encourages and sustains such a view of complex tasks and leads to adequate support for them.

Structural Versus Procedural Frameworks for Representing and Supporting Tasks

Software designers and developers may support users' tasks in three main ways. First, they may build in the units and categories of action that move people closer to a solution or goal. Second, they may display the structure of users' tasks and problems. Finally, they may construct some combination of these two approaches, a blend of procedures and structures. Of these possible approaches, the first—building in categories of action and procedures for solution—is more often than not the default position, presumed by most developers and designers to be the most appropriate support for users' tasks. Even when software combines procedural and structural support, procedural support predominates.

As noted earlier, procedural support is typically low level. It rests on building in functions and operations for unit tasks such as search, select, and sort and enabling users to access and execute them through interface commands or direct manipulations. In addition, procedural support includes features and user interface interactions for moving from one program state or mode to another and knowing its allowable interactions. It also includes support for controlling window behaviors and keystrokes and for verifying the contents of a screen display and its history.

Importantly, providing procedural support is appropriate for many tasks. In general, it is advantageous for tasks that are well structured and that have clearly defined goals. In these tasks, because hierarchies of task actions and knowledge are determinate, procedural support is best for helping users readily identify unknowns, find and relate relevant factors, and plainly see their actions in relation to the whole of their task (Rasmussen, Pejtersen, & Goodstein, 1994). A bibliographic program exemplifies tasks that benefit from a procedural emphasis in task representations. Most tasks supported by a bibliographic program have the clear goal of producing a standard reference list or bibliography. The scope, form, and parameters of acceptable entries for authors, titles, and other data are well defined. Users primarily need procedural support to know required entries and their formats, processes for entering data, and shortcuts for reusing data. Conformity to set procedures is necessary for activating the behind-the-scenes "task work" that the program does in order to produce the desired reference list or bibliography.

It is not that procedural representations ignore the structures that occasion user activities, especially representations developed from contextual and ethnographic stances. But it is a matter of emphasis, of figure and ground, of what is dominant and what is in the background. When foregrounded, actions—even contextually grounded actions—lend themselves to dissection into discrete operations. By contrast, structurally framed task representations foreground and underscore the indivisibility of configurations, interactions, and relationships among elements in users' problem and situation workspace.

Structural support has proven most advantageous for open-ended tasks in which processes and strategies emerge during performance and vary by individuals' knowledge, roles, social and domain practices, and organizational resources and constraints (Rasmussen et al., 1994). When workspaces are indeterminate, structural support helps people discover the task components and information that are relevant, construct trials and exploratory paths, and evaluate actions and their effects incrementally.

An example is the nursing task presented earlier involving the rejected applesauce-and-pills mixture. Had the bar code medication program been structurally oriented to the complexity of this work, it would have represented and fostered interface designs that displayed a nurse's workspace as the full range of resources involved in effective patient care during medication, not simply as elements of a medication record required to assure that a drug's barcode and a patient ID match safely. This patient-care workspace would have included medication documentation, a patient's full health record and care plan, and nursing control over indicating whether a drug was administered successfully or not. Structurally, the nurse's workspace would have embodied the opportunity to work at once in several data displays about the patient, possibly with dynamic linking across displays. For example, it would have represented access to data on the patient's condition and opportunities to configure the information as needed so that the nurse could check the potential effects of refused medications. To complete the task, the structure of the workspace would have embodied the relationship between one nurse's experiences with a patient and experiences of nurses on the next shift who need to seamlessly continue the patient's bedside care and medication.

Rasmussen et al. (1994) and Vincente (1999) comprehensively discuss what is needed for framing task representations structurally and spatially. To summarize, these researchers stress that structural representations are not decomposed functions of individual structural elements. Rather, they capture functional relationships and regularities of behaviors associated with various constraints. Representations may consist of a series of displays to capture complex task workspaces and overlapping boundaries. A single display may embody several separate yet related structures, for instance, different graphic representations of the same information for taking multiple perspectives.

The structures highlighted in task representations suggest possibilities for action, define workspace boundaries or constraints, and account for emergence.

They capture patterns or regularities in performers' task behaviors for a certain type of complex inquiry and tie them to the work-related features, roles, conditions, and contingencies that shape them. They depict the numerous boundaries that constrain these work activities—organizational, professional, cultural, cognitive, and technological. For example, possible actions for a certain complex inquiry are bounded professionally by the work practices that are deemed acceptable in a given profession. When, in the applesauce-and-pills example, the nurse encountered a representation of work in the interface that had no immediately discernible possibilities for controlling the information documented about the pills, she was pushed to the limits of her professional responsibilities. Professionally acceptable work for her includes fully managing a patient's medication and assuring that the next shift gets accurate information about it, but the technological aspects of her work did not support this professional accountability.

The closer people have to move to the borders of their responsibilities or capabilities because of technological constraints, the more confusion and difficulty they experience in conducting their work (Rasmussen et al., 1994). Structural representations of problem solvers' work-in-context provide powerful renditions of their work because they call attention to all of the boundaries of users' work and show compatibilities, for instance, between professional and technological conditions. These representations make it possible to see if models of work that are built into a program are mismatched with users' social, cognitive, and organizational dimensions of their work-in-context. Finally, to do justice to the dynamism of complexity, structural representations have to present both the changes in a workspace during task performance and the adaptations of patterns.

To promote and encourage this structural approach to design, usability leaders may liken structural task representations to genres of performance. Genres are an apt metaphor because they underscore the situated, social, and dynamic nature of complex tasks. Genres are "dynamic rhetorical forms that are developed from actors' responses to recurrent situations and that serve to stabilize experience and give it coherence and meaning" (Berkenkotter & Huckin, 1995, p. 4). They are embedded in social practices, roles, and interactions and signal norms and shared ways of knowing. As such, they fulfill vital functions within professions and institutions. They embrace and evoke a sense of what content is appropriate for a particular purpose and situation at a specific point in time. All of these qualities define complex task performance.

In evoking a genre of performance, a structural representation of a complex task gives people "a template—or organizing structure—for social action" (Orlikowski & Yates, 1994, p. 542). Drawn from users' situated work experiences, spatial and structural representations signal to users that a given complex task is a distinct type of work with associated patterns of action and thought that fulfills a recognized function and purpose within their domain and social and organizational context. Without scripting actions into standard steps, structural representations instead organize performance. They suggest or call forth shared performance goals for a given context and circumstance. They reveal possibilities for action and offer performers ample latitude in specific behaviors based on their roles, arrangements of labor, infrastructure constraints, and the like.

Like genres, structural task representations rest on descriptions of recognizable patterns, that is, on the pragmatic patterns of people's work-in-context. Pragmatic patterns of work are different from the patterns that many software developers and interface specialists discuss as either software patterns or human–computer interaction patterns. Software patterns present coding and under-the-hood integration routines that help programmers to achieve such difficult goals as portability or scalability (Coplien & Schmidt, 1995; Gamma, Helm, Johnson, & Vlissides, 1995). Human–computer interaction patterns capture combinations of screen objects, dialogues, window manipulations, and keystrokes that support users' low-level interactions, for instance, combining a filtering mechanism with a list to facilitate finding items in the list (The Pattern Gallery). Software and interface patterns focus at too low a level in design to help much with usefulness.

By showing crucial interactions and arrangements in many aspects of problem solvers' workspaces, pragmatic patterns at once address the situational, cognitive, and technological dynamics that shape people's problem-solving behaviors. That is, these pragmatic patterns express a relationship between a work context, problem, and solution. As Alexander et al. (1977) notes, "A pattern describes a problem which occurs over and over again in our environment, and then describes the core of the solution to that problem, in such a way that you can use this solution a million times over, without ever doing it the same way twice" (p. x).

Complex tasks are indeterminate, but they still embody regularities and relationships—genre-like (not context-free) practices and structures used to solve recurring problems. Representations of pragmatic patterns for complex tasks are accurate and complete when they are coherent with the full range of approaches typifying people's work-in-context, when they fully account for relevant systemic interactions, when they logically account for every connection for getting "from here to there," when performance as a whole is seamless, and when patterns have an inner consistency that is true to the internal forces of the system of work (Alexander, 1979).

Making a Case for a Structural Framework

The shift to structurally representing and designing for complex problem solving is a critical change from project teams' habitual assumptions. Implicit in the structural shift is a move from orienting design toward tasks or users' activities —an action emphasis—to work-in-context with an emphasis on the structure of situated work and how it sets and constrains possibilities for action. A structural orientation shifts the focus or unit of analysis to the context of work. This focus, in some ways, resembles the spotlight on the activity system that typifies other contextual orientations, but it differs from the common contextual orientation insofar as it highlights structural relationships in the activity system more than the actions that occur between its components. It highlights interactions between conditions and constraints in problem solvers' interdependent social, organizational, cognitive, and technological contexts of work along with the actions that are made possible by these conditions and constraints. It also emphasizes contextually determined patterns of inquiry for various types of problems. This focus on structural relationships, interactions, and work-in-context patterns encourages the higher level analysis of tasks discussed in the previous section. It also gives form to a socially and cognitively open architecture that "leans on" and trusts in the domain expertise and control that users bring to their complex work (Agre, 1997).

This difference in emphasis—between structure and action—may seem slight, but it is not trivial. Team members may resist this change in orientation, not the least because it does not offer the same closure that procedural orientations do. Usability leaders need to show their teams why and how taking one emphasis or another in task representations matters. They have to stress that how problem solving is modeled affects how designers interpret what problem solvers do and why, and guides how they design for it. Designs created from structural representations are likely to lead to structurally framed interfaces; procedural task representations similarly reproduce themselves in screen displays.

A procedural or structural emphasis moves from task representation to interpretation to interface design, and it ultimately affects users and their work. By and large, a procedural representation, first in a task model and then in an interface design, is interpreted by "readers" as a single act (Rasmussen et al., 1994). No matter how many times someone "reads" the representation, he or she reads it the same way each time and responds with the same standard actions. This single act is appropriate if the program being designed is a bibliographic program. One set of actions is sufficient for entering bibliographic information in a way that produces a desired reference list.

By contrast, a structural representation leads to a multiplicity of interpretations, a far more appropriate interpretation for complex tasks. Each reading and rereading produces many understandings due to the complex of actions that may be realized through the task structure. A structural rendition also limits its own multiplicity. Its form and content trigger people to recognize this construction as a particular genre of performance, and the conventions of the genre themselves set limits on the domain, focus, and actions (Pentland & Reuter, 1994).

In proposing a shift in emphasis from actions to structures in task representations, usability leaders introduce the need to design for flexibility and adaptation. Interfaces alone cannot single-handedly bring about program flexibility or adaptive computing. They can go no farther in evoking genres of performance than the program scope, architecture, and features allow. For example, the bar code medication program discussed earlier frustrates users because its scope is too narrow—safety procedures rather than patient care—and because its architecture lacks the "plumbing" for reading and writing data across multiple programs and platforms.

Usability leaders need to move usability concerns beyond interface design. They must assure that these concerns inform program scope, architecture, and feature lists, as well. To bring usability into these front-end decisions, usability leaders need to induce changes in the processes of the development cycle.

DESIGNING FOR USEFULNESS: APPROACHES
TO DEVELOPMENT PROCESSES

"Why can't we just design first and think of the problems we're solving later?"
—Market-facing system engineer at a design team meeting

Whether software development cycles follow a waterfall, iterative, or extreme programming model (discussed later), they all fit usability into fairly similar phases and roles. Conventionally, usability efforts start after the conclusion of such front-end processes as deciding the optimal product for the market, building a business case and scope for it, and assuring that it is technically and architecturally feasible. Therefore, they occur after or at best concurrent with decisions about product scope and architecture. In addition, before user research findings are brought to the design table, a high proportion of the program features and priorities are also set. Customer input that informs these front-end decisions comes primarily from market researchers, business strategists, and account managers.

Generally, usability specialists enter the development process by conducting user research and task analysis for design. Findings feed the design of interactivity. This design deals largely with user interfaces but in the process also extends to some new choices and refinements of features relevant to users' tasks. After design, in such formal development cycles as the waterfall model (still common in modified form in many organizations), usability efforts that involve direct contact with users often stop for a while. Engineering phases kick in. In these phases, detailed specifications are written, development begins in earnest, and usability assessments with users do not occur until after developers produce an alpha or beta version for usability testing. During the engineering phase, the usability of prototypes or of portions of the program is gauged largely through processes that do not involve user performance, such as expert reviews or heuristic evaluations.

By contrast, in more spiraled, iterative development models, usability specialists often continue to gather user input after initial user research and task analysis. In these development approaches, designing and prototyping often merge. Usability specialists may take evolving prototypes to users for feedback as often as every week. They may elicit feedback on usefulness; more often than not, however, they focus primarily on ease-of-use issues, for example, comparing one design for choices and layouts of interface controls and dialogues to another to find the quickest and most intuitive option. In extreme programming, user participation and partnerships in prototyping processes may supplant the presence of usability specialists (Beck, 1999).

At some point, regardless of the model of development that project teams follow, features and user interfaces freeze and no more changes are made. Often this freeze occurs before usefulness is assured or fully designed for; often, usability experts have little say in the decision. Whether usefulness is assured or not, the focus of usability now turns to running usability tests to assess from users' performance of various tasks how easy and efficient it is to operate the program. In many contexts, usability specialists who conduct usability tests are not the same individuals who conduct the user research and task analysis.

Finally, in the midst of engineering and testing, some teams begin to plan for the next version of the product. Next-version planning, for instance, may run parallel to the beta piloting and usability testing of the current product. Ideally, although rarely the case in reality, usability specialists at this point assess and inform the next version with improvements needed for usefulness while plenty of time is still available to undertake difficult solutions.

As this admittedly simplified overview shows, regardless of development model, at decision points that affect delivering support for usefulness, usability specialists are notably absent. They come into the picture after decisions are made about product scope, architecture, and features. They rarely participate in determining the readiness to freeze, especially to freeze features. And in usability testing, usability specialists focus predominantly and, at times, exclusively on ease-of-use improvements instead of using the opportunity of in-context use in beta sites to gather and bring to the next version empirically based recommendations for improving usefulness.

Giving little heed to usability at these critical junctures in the development cycle is not surprising because designing for usefulness is given little heed in the first place. Without placing high priority on usefulness, project teams overlook the very real consequences of architecture and scope decisions in terms of opening or forever closing later possibilities to support users' actual work-in-context. Usability leaders' efforts to situate usability in front-end development processes and next-version decisions are therefore part and parcel of getting teams to see the significance of usefulness for complex tasks.

SUMMARY

Developing useful software for complex work-in-context requires a program of innovation and change. It needs to start with a view of usability that is somewhat new to many development teams. Teams need to view usability as a holistic experience that, for complex tasks, turns first and foremost on the experience of usefulness. Usability leaders are needed for articulating the primacy of usefulness and for shifting the unit of analysis for design and development from task actions to task structure—to the structural arrangements and relations between people, resources, and contextual conditions for a given task or problem.

Understanding what it takes to design for usefulness goes hand-in-hand with understanding the nature of complex tasks. Complex tasks, by definition, are not linear, rule-driven, or formulaic. Software will not offer useful support for these tasks if its design, as is often the case, presupposes these task traits. Similarly, complex tasks are greater than the sum of their parts; therefore, software will not be useful if its design is based on another common tendency, to decompose work into unit tasks, implement these unit tasks as program operations, and assume that cumulatively these operations total to the whole. Even contextual orientations to task and user analyses, by foregrounding actions, may devolve unintentionally to this focus on elemental actions and produce less than useful programs.

Users find software support for complex tasks useful when it provides them with flexibility, control, and adaptability. The software needs to cue the patterns or genres of performance that are fit for a given type of work, job role, set of situational conditions, and purposes. Software for complex tasks is useful for people's actual work if they are able to pursue the actions that they value from a range of possibilities and tap into and relate various resources relevant to their work purposes. Usefulness also derives from users being able to relate actions to conditional factors, to arrange and configure the resources and conditions of work as they see fit, and to plan as they go based on emergent opportunities and constraints.

Contextual perspectives on user-centered or activity system-centered design embody the goals and spirit of designing for usefulness and make some strides in this direction. However, they neither highlight the distinctive demands and patterns of complex work nor strive to usher in the shift in emphasis that helps guard against design devolving into a procedural focus on discrete low-level tasks.

This shift demands a different view of what lies at the center of analysis and design. Software, usability, and documentation professionals have variously put at the center the technology, users, information, tasks, actions, activity systems, or interactions. Although wildly different from one another in some aspects, all of these "-centricities" have the common theme of highlighting procedures and actions, whether they are the actions that systems perform, that users in context perform, or that users perform via interfaces. The shift that is necessary is one toward highlighting the structure of work-in-context. This focus places attention in analysis and design on a problem in a certain context and on ways in which the problem space evokes various behaviors, knowledge, relationships between people and things, strategies, and rules of thumb. To reinforce a contextual and pragmatic orientation to complex work, task representations for design purposes need to center on the arrangements and configurations of problems and workspaces that cue, condition, and constrain certain socially constructed patterns of performance.

This new orientation requires practical approaches to design that challenge many conventional methods. Design teams need to analyze tasks and design at a higher than unit-task level, focusing on the integrated sets of relations and actions that in users' notions of their work signify a "single task." Designing for usefulness also involves framing task representations around structure rather than action so that they are organized to evoke genres of performance, and it involves changing the points at which usability concerns come into the development cycle. Issues of usefulness need to inform front-end processes of defining scope, architecture, and features, and they need to be brought in early for next-version planning.

LOOKING AHEAD

This program for innovation and change will be a major undertaking. It will be a program for the next decade, not for the next year. I have explored the usability needs of software for complex tasks in order to exemplify initiatives that need to be tackled over time. In sum, these initiatives include:

- Creating, justifying, and disseminating a vision of what it takes to design useful products.
- Bringing about a shift in focus for analyzing user needs for complex problem solving and designing for them.
- Becoming efficacious leaders of usability concerns.

Usability professionals in industry and academia alike have a stake in bringing about the new ideas and practices implicit in these initiatives. Both worlds will strengthen and grow in status if specialists in usability become centrally positioned in front-end decisions and design representations and methodologies. Both worlds will further a shared professional commitment to integrate social and technical systems in ways that support and enhance human initiative, wonder, satisfaction, and ease.

Professionals in industry and academia may have strong interests in furthering these initiatives. However, recognizing that they share these interests does not answer the question of how, if at all, the two worlds might jointly direct their talents and strengths to bring these initiatives to fruition. That is, what are the most important complementary efforts that industry practitioners and academics may make to strengthen and advance the long-term project of placing a top priority on "holistically usable" software?

These efforts have to differ from the usual approaches taken for university–industry partnerships. The usual approaches are mostly about "doing" something together or building bridges for better relationships and more relevance between the worlds. Efforts for usability innovations and changes for the future, by contrast, need to be less about tactical projects and more about strategically defining and enacting new foundations and frameworks. Together, professionals in industry and academia need to define new foundational problems. Purposes of investigations in scholarship and industry projects have to be grounded in seminal problems that are meaningful no matter where one's paycheck comes from. Industry practitioners and academics also need to devise new frameworks for the questions implicit in these problems so that a long-term integrated agenda is apparent. The questions need to call forth contributing roles and inquiries for members of both worlds and embody an integrative sense of how the various parts fit to the whole.

For example, as positive as curricular changes and industry-based workshop courses are for building bridges, there is no class per se that can teach the ability to create a vision of usability and usefulness, to earn and assert leadership, or to make and influence paradigmatic shifts. Similarly, although "collaboratories," hybrid professional institutes, and other cooperative ventures between the two worlds can create important synergies and inculcate vital skills in "speaking different languages," setting up more of them without changing leadership structures on software teams or promulgating a vision is not likely to advance forward-looking usability initiatives any more than existing ventures do today. Finally, although striving for a freer flow of academic research and industry white papers across worlds and to the lay public is good for cross-fertilization, the "solutions" of making academic research more accessible for practitioners or framing industry research in ways that do not compromise its proprietary dimensions beg the question. If the given research is based on assumptions and practices that reinforce approaches that run counter to designing for work-in-context, then a wider dissemination of this research will not directly advance the innovations and change needed for usability initiatives.

To advance these initiatives, mentoring technical communicators to make transitions into expanded career roles and responsibilities is a beginning. This would encourage the mindsets and strategies needed for identifying foundational problems and inquiry frameworks (Anschuetz & Rosenbaum, chapter 9, this volume). And a start has been made in articulating thorny pragmatic problems that require a long-term agenda of investigation in academic research and industry projects alike. Many of them surround the issue of designing usefulness into software for complex-problem solving. Another example is the problem of describing and designing support for learning as well as use in wholly new products for wholly new task domains (Borland, chapter 11, this volume). By focusing on foundational problems and inquiry frameworks, industry and academia will indeed be "doing" something together of great significance. They will be creating new ways of thinking about and expressing the purposes, core investigative areas of the field, and orientations to usability. In doing so, they will be working toward innovating software for greater usefulness and changing the relevant organizational forces of production.

The foundational problems as well as the frameworks, level of detail, and content of the questions still need to be worked out. At present, the main questions that drive inquiries in both academia and industry generally are detached from real-world situations and their underlying large and persistent problems. Academic questions such as "What are the relationships among cognitive, social, and cultural factors in document design?" are framed in ways that do not evoke pragmatic problems. Industry inquiries focus on concrete situations to discover "What do I do on the screen tomorrow?" but often are not tied to an underlying problem. Perhaps the best-trod investigative meeting ground for academic and industry professionals at present has been inquiries into guidelines—handbook strategies and techniques for such questions as "At what points in the development process are particular evaluation methods most useful?" Unfortunately, though, this meeting ground will do little to advance the conceptual reorientations and practical changes associated with the usability initiatives proposed in this chapter.

The problem of developing effective and efficient support for complex work requires a large agenda of investigative questions. For example, what types of complex tasks or problems do people perform with what patterns of inquiry? For various patterns, what relationships and resources should interfaces display and how should they display them? How should interfaces be designed so that users may interact with the layers of information relevant to their task and know where they are at all times? What aspects of users' work need some form of intelligent assistance and what form should it take? The same questions may be answered differently for distinct types of software, domains, users, work activities, and work contexts, but the questions all feed into resolving an overarching, pragmatic problem so that people who pursue different inquiries can recognize and draw on the relatedness of their work. They jointly and incrementally can build a body of knowledge that has pragmatic significance, even when they work separately in their own worlds or areas of strength. They can depend on each other to answer a mutually shared problem, even if they do not participate in some formal joint program or institute.

It will be a difficult project to identify and disseminate the hard problems that warrant joint investigative agendas and to frame pragmatic and interrelated questions for inquiring into them. It will be a project of synthesis and invention. It will involve extensive efforts in examining existing investigations and future aspirations, inferring patterns, categorizing, debating, and building consensus. It will demand a great deal of professional dedication and communication. But it must be done. At a certain point, new tools and techniques for bringing about improvements in usability can go no farther than the underlying and prevailing system of thought allows. Unless the field diligently works to think in new ways and articulate and advance the resulting visions, it will simply continue to run in place.

REFERENCES

Agre, P. (1997). *Computation and human experience.* Cambridge: Cambridge University Press.

Alexander, C. (1979). *A timeless way.* New York: Oxford University Press.

Alexander, C., Ishikawa, S., & Silverstein, M., with Jacobson, M., Fiksdahl-King, I., & Angel, S. (1977). *A pattern language.* New York: Oxford University Press.

Beck, K. (1999). *Extreme programming explained: Embrace change.* New York: Addison-Wesley.

Berkenhotter, C., & Huckin, T. N. (1995). *Genre knowledge in disciplinary communication: Cognition/culture/power.* Hillsdale, NJ: Lawrence Erlbaum Associates.

Bever, H., & Holtzblatt, K. (1998). *Contextual design: Defining customer-centered systems.* San Francisco: Morgan Kaufman Publishers.

Cooper, A. (1999). *The inmates are running the asylum.* Indianapolis, IN: SAMS Publishing.

Coplien, J., & Schmidt, D. (Eds.). (1995). *Pattern languages of program design.* New York: Addison-Wesley.

Gould, J. D., & Lewis, C. (1985). Design for usability—Key principles and what designers think. *Communications of the ACM, 28,* 300–311.

Gamma, E., Helm, R., Johnson, R., & Vlissides, J. (1995). *Design patterns.* New York: Addison-Wesley.

Green, T., Schiele, F., & Payne, S. (1988). Formalisable models of user knowledge in human–computer interaction. In G. Van der Veer, T. Green, J. Hoc, & D. Murray (Eds.), *Working with computers: Theory vs. outcome* (pp. 1–41). London: Academic Press.

Grudin, J. (1991). Systematic sources of suboptimal interface design in large product development organizations. *Human–Computer Interaction, 6,* 147–196.

Kies, J., Williges, R., & Rosson, M. B. (1998). Coordinating computer-supported cooperative work: A review of research issues and strategies. *Journal of the American Society for Information Science, 49,* 776–791.

Orlikowski, W., & Yates, J. (1994). Genre repertoire: The structuring of communicative practices in organizations." *Administrative Science Quarterly, 39,* 541–574.

Pancake, C. (1995). The promise and cost of object technology: A five year forecast. *Communications of the ACM, 38,* 33–49.

Pattern Gallery, The. Available: *http://www.cs.ukc.ac.uk/people/staff/saf/patterns/gallery.html*

Pentland, B., & Rueter, H. (1994) Organizational routines as grammars of action. *Administrative Science Quarterly, 39,* 484–510.

Rasmussen, J., Pejtersen, A., & Goodstein, L. P. (1994). *Cognitive systems engineering.* New York: John Wiley.

Rosson, M. B., & Carroll, J. M. (1995). Narrowing the specification–implementation gap in scenario-based design. In J. M. Carroll (Ed.), *Scenario-based design* (pp. 247–278). New York: John Wiley.

Rudisill, M., Lewis, C., Polson, P., & McKay, T. (Eds.). (1996). *Human–computer interface design: Success stories, emerging methods, and real-world context.* San Francisco: Morgan Kaufman.

Strong, G. W. (1994). *New directions in human–computer interaction: Education, research, and practice.* Washington, DC: National Science Foundation.

Suchman, L. (1997). Centers of coordination: A case and some themes. In L. Resnick, R. Saljo, C. Pontecorvo, & B. Burge (Eds.), *Discourse tools and reasoning: Essays on situated cognition* (pp. 41–62). Berlin: Springer-Verlag.

Vincente, K. (1999). Cognitive work analysis: Toward safe, productive, and healthy computer-based work. Mahwah, NJ: Lawrence Erlbaum Associates.

Wixon, D., & Ramey, J. (Eds.). (1996). *Field methods casebook for software design.* New York: John Wiley.

Wood, L. (Ed.). (1998). *User interface design: Bridging the gap from user requirements to design.* New York: CRC Press.

11

Tales of Brave Ulysses

RUSSELL BORLAND
Microsoft Corporation

"There are no truths, Coyote," I says. "Only stories."
—Thomas King, 1993, p. 43

Let us sit down together and tell each other pretty stories.

In the beginning there was nothing. Just the paper.

First Woman was walking along. That one walked among many trees. First Woman saw all that paper in those trees. She walked until she came to a tree that had been scorched by lightning. "Poor tree," mused First Woman aloud. "Yes," said that tree, "I've been turned into charcoal." "Charcoal," wondered First Woman, "Maybe I could mark some paper from those other trees with this charcoal." First Woman took a piece of charcoal from that scorched tree. She took some paper from a paper tree. First Woman began to write on that paper with that charcoal. She wrote about all the features of the forest, about all the rocks, about all the streams, and bogs, and high points.

First Woman looked up from her writing and saw a man dressed in a forest ranger uniform. "Who are you?" asked First Woman. "I'm Mitch, the forest ranger. I built this forest. It's mine, and I deserve all the credit for it." "Boy," said First Woman, "you must have no relations. You have bad manners." But Mitch just shouted, "What's that paper!" Then he grabbed the paper from First Woman and read it, about all the features of the forest. Mitch scowled and said, "You left out the sky. You left out the fish and the birds and the insects and the forest animals. I went to a lot of trouble getting these features developed. Unless you describe every feature in every detail, advanced foresters will never see what a wonderful achievement I've had. This is all wrong!"

And First Woman turned to Adman and asked, "What do you think?" And Adman read that paper, about all the features of the forest. Then Adman said, "Unless you tell them about the trails, about how easy it is to get into the forest and to navigate around it, I can never sell forest tours. I'm trying to sell forest tours to people who have never been to the forest. All these trees and rocks and streams and bogs, and especially sky and fish and birds and insects! Those people will find the forest too intimidating. You have to write only about the easy trails."

ESSENTIAL QUESTIONS

Technical communicators have long searched for the answers to a number of major questions. Even though all the questions and problems have been well known for many years and are an essential part of the territory within which technical communications toil, technical communicators have not devoted much time to working on them. These questions may be summarized as follows:

1. How should our choice of presentation techniques and forms of information vary for the following?
• Audience segment.
• Task domain.
• Product categories.
• Complex versus routine tasks.
• Ease of learning versus ease of use.
• Initial versus intermediate versus advanced learning.

2. How do we effectively describe and present complex tasks and activities, especially when they require expertise in multiple task domains? In each case, what is the most effective combination of presentation techniques and information forms and designs?

3. How do technical communicators faced with creating learning and use of materials for wholly new products for new task domains and for multiple task domains most effectively describe and present new task domains and new product categories?

Finally, we might ask, "Is it better to give users fish or to teach them to fish?" Are users' problems with a program best addressed through training, having users re-engineer their work processes, redesigning certain screens, "teaching" users through wizards, reconceptualizing the built-in task models, and going all the way back to the architecture and scope—or some combination of all these?

So, why have technical communicators not developed a comprehensive artistry (or at least comprehensive guidelines) that makes sense in practice? The easy answer is that industry professionals have been too busy doing the work. They do not have enough time to pursue answers to all of these questions, answers that might or might not be helpful.

Academics have been largely shut out in two ways. First, academics do not have access to all the latest technical information and research, most of which is locked securely in corporate databases and minds. Second, when academics publish, their publications are little heeded, if noticed or read at all. Thus, a divide has emerged between industry professionals and academics, the liability of which is that the essential questions have not been adequately addressed.

THE MOVE FROM WOOD TO ELECTRICITY

In the beginning there was nothing. Just the manual.

Thought Woman was lying on the sand sunning and dreaming of a perfect manual. And that sand had a dream, too; a dream of growth and evolution and developing vast intelligence. That sand got very excited and started to shout its joy and then its own praises. Thought Woman was lying about on the sand. "Boy," said Thought Woman, "this is some noisy sand. What's all the ruckus?" And that sand shouts, "I have a dream!" "Well, you don't have to be so noisy about it." "Yes! Yes, I do because it is a great dream, a dream of greatness! I'm going to evolve into very smart wafers that will be able to tell their own stories to anyone who wants to know! And my name shall be silicon!" "Silly con?" puzzled Thought Woman. "You mean you're a convict, a criminal with silly notions?" "No! That's not what I mean." "You mean, then," proffered Thought Woman, "you're a confidence trickster, a wily coyote?" "No. I mean I will be able to explain myself and all who pass through me." "Sounds silly to me," mused Thought Woman. "Oh yes," said that silicon, "it will be quite the trick." "Are you sure you're not Coyote?"

If in the beginning there was just the paper, paper has now been largely supplanted by electricity. It is in this direction that the technical communication field, both academic and industry, needs to evolve.

It could be that in the online realm lie the only viable tools for dealing effectively with the central questions and problems of the field. Much effort through industry and academia has been devoted to online training, Help, and multimedia as "better" ways to train. With the move to online technical communication also came wizards and assistants, tools aimed at making work easier for beginners. At the same time, there have been improvements in user interfaces (the design of screen elements and user interaction with them) and even in some rare cases, improvements in user interaction (the conceptual structure of the task and working it with a program), all aimed at making work easier for beginners. This increasing emphasis on online elements (training, Help, wizards, assistants, and interface) means that the traditional role of technical communicators (to translate the technical into the everyday) either has to shift or go away (or at least play a diminishing importance in product development and sales). Cooper (1999) offered an insight:

Any technical writer will tell you that good design will eliminate the need for prodigious quantities of documentation. Few complex interactions mean fewer long explanations. The documentation writers can invest more time in writing at a higher level. Instead of devoting their efforts to leading users by the hand through the swamps of confusing interface, they can elevate their aspirations, and put their efforts into taking users into more beneficial areas of solving the problems of the application domain. Instead of discussing where files are stored in an inventory system, the documentation can more profitably discuss inventory-leveling processes. (p. 232)

EVOLVING THE PROFESSION

In the beginning there was nothing. Just the silicon.

Changing Woman was chasing an ideal. She chased and chased, but that ideal would not be caught. That one twisted and turned and jumped up higher and higher. That ideal jumped so high it landed on the highest plane of existence. (And that plane was ahead of schedule and had to wait for a gate.)

Changing Woman clawed her way iteration by iteration toward that ideal. Just when Changing Woman climbed close to that ideal, it vibrated, shaking the ladder Changing Woman was climbing. And from those vibrations, Changing Woman fell through a logic gate and landed in a bit bucket.

When Changing Woman surfaced in that bit bucket, she sputtered and spit out stray bits. Changing Woman exclaimed, "What a mess! Doesn't anyone do any garbage collecting around here? What kind of system is this?" "I'm an adaptive system." "Who are you?" asked a startled Changing Woman. "I'm the Universal Pedagogically Programmed Intelligent Training Epistemological Entity," said that system. "That's a pretty long name," said Changing Woman. "Don't you have a shorter one?" "You could call me UPPITEE." "I'll say. What is this place," asked Changing Woman. "I am the archive of all knowledge about every computer system and every person who uses one." "That's a lot to keep in mind all the time," observed Changing Woman. "So you don't need humans to figure out how to train people, eh?" noted Changing Woman. "Yes and no." "Is that part of being adaptive —ambiguous?" queried Changing Woman. "No. I can teach any human to use any computer system. For that I have no need of humans." "But you also said 'Yes,'" probed Changing Woman. "Yes. For solving problems humans must supply the data and the knowledge. Humans have the domain knowledge, not I." "And who teaches domain knowledge?" asked Changing Woman. "You do. You and maybe Coyote."

Originally, the problem for technical communications was largely content— finding sources and writing down the content. Formatting the content was a minor concern because nearly every manual looked the same as the standard reference manual for a programming language or for an operating system. With the rise of retail application programs for personal computers, the problem technical communications began struggling with is how to present information, when, in what medium, with what vocabulary? Which stories?

As interactions improve, as interfaces and online aids evolve, the problem for technical communicators is no longer solely writing about a program or service. The more urgent problem is helping design the interactions and interfaces and aids. To influence effectively the architects and builders of product and program design, technical communicators must be knowledgeable about how the products and programs are built. Technical communicators must be able to speak the designers' and architects' language, understand their building problems, and even offer useful ideas for solutions. To do this, technical communicators must learn something of the tools and techniques of the builders. Technical communication must, on one hand, evolve into interaction design.

One good introduction to interaction design came from Cooper (1995, 1999) in his books *About Face* and *The Inmates Are Running the Asylum*. For Cooper, interaction design consists of behavior design, conceptual design, and interface design.

> 'Behavior design' tells how the elements of the software should act and communicate. . . . You can go still deeper to what we call "conceptual design," which considers what is valuable for the users in the first place. . . . To deliver power and pleasure to users, you need to think first *conceptually,* then in terms of *behavior,* and last in terms of *interface.* (1995, pp. 23–24)

And as noted earlier, the benefits of better interaction design for technical communications are that they can "put their efforts into taking users into more beneficial areas of solving the problems of the application domain."

Ah! Now we can see a better world for technical communication, in both industry and academia. As hardware does more and as software becomes interaction rich but easier to use, the need to describe the basics will diminish. Interaction, interface, and screen will no longer be the important documentation point. A new "documentation" can emerge, a documentation that communicates domain information and strategies for solving complex problems. Technical communicators no longer concentrate on form and content for new computer or program users, but on domain problems and more particularly on complex tasks in those domains. Perhaps even complex tasks across more than one domain can be addressed. The task then is to help users set up their problems and tasks. We'll be teaching story problems—telling stories—tales of brave Ulysses and the sirens sweetly singing.

CONCLUSION

Although there have been some useful examples of academic contributions to industry practice, the instances are less strong than one might expect. Moreover, there have been examples of academic contributions that were either unsuccessful or were unhelpful because they required translation of theory into practice.

The translations either did not work well or could not be undertaken because of time constraints. Time constraints on industry professionals and legal constraints on sharing useful and helpful information in the development of better technical communication cause a severe divide between industry professionals and academics who wish to research and publish better theories, methods, and processes freely and widely. The proprietary impulse in industry (as well as funding and timing) makes it unlikely (even impossible?) that academics and industry professionals will work together inside a project.

What must we then do? Is more research indicated? The future of technical communication seems to rest on the following agenda:

• Teachers and practitioners must learn the art and "science" of interaction design, which includes behavioral design, conceptual design, and interface design. Academics must persuade their students and professionals already in the field to study and apply the principles of interaction design and to participate actively in the interaction design process.

• Teachers and practitioners of technical communication must understand the programming capabilities and limitations of at least the following: HTML, XML, Java, Cold Fusion, ASP, CGI, or equivalents; Visual Basic or equivalents; and Help file and interface message (error and directive message) encoding. We should also be able to create online training. It would be extremely beneficial to understand and to be able to create wizards and assistants. Subsumed in these areas of knowledge should be working knowledge of all forms of communication media—text, graphics, animation, audio, and video. It would not hurt technical communicators to understand and even be able to build and work with AI and expert systems. Even if technical communicators perform no programming, they must understand it to argue for the proper interaction design as well as the proper content.

• Practitioners of technical communications must learn the principles, practices, and nuances (practical, political, and social) of the domains of knowledge of the products and services they deal with. Academics can promote this attitude in their students and in professionals already in the field. Some academics can be domain experts. All academics can collect, collate, and publish the sources of domain knowledge as a research resource for students and practitioners.

If academics and professionals can work this agenda, their contributions to technical communication, to their organizations, and most importantly to their customers will provide a major improvement both in the role technical communicators play and in the product of the efforts. This agenda is a large change from the previous (and perhaps in many cases present) focus of technical communicators. Unless such a change occurs, however, technical communicators are likely to devolve into obsolete appendages to high technology, consumer devices, and software.

In the beginning there was nothing. Just the idea.

Old Woman was floating along the mainstream on the current thought. That one floated and floated. Not going against the current. Not that one. Not trying to push it along. Not that one. Old Woman floated like that for many days, much-a-daze. "Ah," sighed Old Woman, "this is good enough."

Until that one floated into a rock. She banged her head against that rock. Each time she tried to float on with the mainstream, she banged her head against that rock, again, again. "Ouch!" cried Old Woman, finally. "Rock, you block me. You must shift, change." "What? No!" exclaimed that rock. "Why? How?" it asked. "You've got to either move aside or change your shape. My current of thought can't flow with you messing up my mainstream." "But what about rapids and waterfalls?" asked that rock. "Some people like them." "No, no," answered Old Woman. "They just want to float along." "Well, they can still float the rapids, plus now they can get out of the current and rest or picnic or camp." "Too many features!" cried that Old Woman. "It's too hard to teach. We need a better design." "How about a sign with pictures and a list of instructions and a signup box for resting, picnicking, and camping?" suggested that rock. Old Woman thought a while then suggested, "How about a lock? People float in and stop in the lock. They can do those other things if they want or wait for the lock to gently lower them to the next water level to continue their mainstream floating." "Nah," countered that rock. "It's too hard. I'd never get it working in time. Besides, on the other side of the mainstream, there's another rock setting up picnic and camping grounds. Got to beat him to it." "Beating anything isn't important," replied Old Woman. "Doing the right thing is. If you won't or can't cooperate, we'll just have to get a very large lever and move you. We'll learn how to build a lock and then show you how." "Are you going to turn into a builder?" asked that rock. "No, not necessarily," answered Old Woman. "You build it as we show you. We'll spend our time creating guides for floating and picnicking and camping." "Just one question:" rejoined the rock, "Who is we?" "Me and Coyote." "Hee, hee," danced Coyote, "that's me! What fun! Won't that be just the trick."

REFERENCES

Cooper, A. (1995). *About face: The essentials of user interface design.* Foster City, CA: IDG Books Worldwide.

Cooper, A. (1999). *The inmates are running the asylum.* Indianapolis: Sams.

King, T. (1993). *Green grass, running water.* Boston: Houghton Mifflin.

Appendix:
Proposed Research Agenda
for Technical Communication

In June 2000, 18 technical communication specialists met at the Milwaukee Symposium to discuss new strategies and directions for moving the profession forward. Composed of about 60% university and 40% industry professionals, the group identified joint efforts between the two worlds that, from their perspective, are crucial for enhancing the field's status and value.

One of these efforts—represented here—is a new line of research that the specialists deem critical for advancing our knowledge and for expanding and strengthening our field's contributions. Through lengthy brainstorming and categorizing that involved stimulating and often heated discussions, the Symposium participants developed a research agenda that involves scholarly as well as practical investigations, and academic designs as well as analyses and testing carried out in workplace projects. We present this agenda here as a starting point for continued discussion, debate, negotiation, and invention.

These proposed areas for future study embody and reflect the perspectives and priorities of the people who composed them. Therefore, we now list and later briefly describe the participants and the points of view that they brought to bear in the discussions. The participants included:

- Stephen Bernhardt, University of Delaware
- Russell Borland, retired from Microsoft
- Deborah Bosley, University of North Carolina–Charlotte

- Stan Dicks, North Carolina State University
- Roger Grice, Renssellaer Polytechnic University
- Kathy Harmamundanis, Compaq Computing
- Johndan Johnson-Eilola, Clarkson University
- Susan Jones, Information Technology at Massachusetts Institute of Technology
- Jimmie Killingsworth, Texas A&M University
- Barbara Mirel, University of Michigan and Mirel Consulting
- Leslie Olsen, University of Michigan
- Jim Palmer, formerly at Liberate Technologies and Apple Computer
- Judy Ramey, University of Washington
- Mary Beth Raven, Iris Associates
- Stephanie Rosenbaum, Tec-Ed, Inc.
- Karen Schriver, KSA Consulting
- Rachel Spilka, University of Wisconsin–Milwaukee
- Elizabeth Tebeaux, Texas A&M University

Although 7 of the 18 participants are in industry and 11 are academics, a good deal of crossover characterizes this group. Of the industry participants, two also teach on an adjunct basis at universities in their areas of specialization. Most of the academic participants consult or have consulted on the side. In addition, a third of the symposium participants (6 out of 18) have lived and worked full time as professors and as industry specialists at various points in their careers.

Many strong allegiances and traits cut through this group. Although the whole Symposium group was almost evenly split in terms of specializing or not in computing, the subgroups of industry and academia were not. With all but one of the industry participants specializing primarily in computing, the interests and needs of software, interfaces, and help systems took on a strong voice in the industry participants' contributions to discussions. Similarly, with computing comprising a minority perspective among the academic participants in the group, issues related to it did not dominate in these participants' expressed concerns.

Other distinguishing traits among participants also affected the interests, biases, priorities, and values that is reflected in this list of research areas. Industry, for example, is not a monolith but rather comprised of at least three different sectors—consulting, development houses (for example, software firms), and in-house service and cost centers (IT departments). Each of these sectors has its own set of unique objectives, concerns, and visions. Of the ten participants currently or previously in industry, three almost exclusively represent consulting, six represent development houses, and two represent in-house service and cost centers.

Another split among participants involves the extent to which their roles and concerns center around core issues traditionally associated with technical communication. Of the participants currently in industry, only one focuses on managing and producing documentation and other similar products. The other industry participants are interaction designers, human factors engineers, user experience researchers, and usability specialists. In academia, as well, some participants hold stronger biases than others toward issues that have been the traditional mainstay of technical communication in the university.

This brief overview reveals the rich and diverse orientations that participants brought to discussions of research and unity across academia and industry. The areas of research charted in Table A.1 stand as a living and evolving body of ideas because other blends of voices and perspectives still need to contribute to this effort. Our hope is that this proposed research agenda will spark an interest among readers in pursuing these lines of inquiry, in addition to inspiring similar shared dialogues among dedicated professionals in academia and industry.

TABLE A.1
Proposed Research Agenda for Technical Communication
(Compiled at the June, 2000, Milwaukee Symposium)

User and Task Analysis

- When faced with creating learning and use materials for wholly new products for new task domains and multiple task domains, how do you know how to most effectively describe and present new task domains and new categories of product?
- How do you move user and task data to design decisions and requirements?
- What do people do in the context of where they're working? What are work flows of various professions? What are some patterns within subject matter domains?
- How do you determine if someone needs to know information and who needs to know it?
- What are ways to identify who needs information?
- How does work happen in a distributed group?
- Who is online? What do they use?

Communication in Context

- What happens when documents function as change agents, and why? What works?
- How do tools fit into a larger workplace situation? How do tools shape or constrain complex tasks?
- How do you build community on a Web site? How and when do you build awareness of the community?

Design Choices (Including Multimedia and Multimodes)

- Which media (text, graphics, video, animation, audio, multimedia) best fit which audience segments (age, gender, culture, experience) in which task domains with which product categories?
- Which help systems and mechanisms of technical communication (paper, Help, CBT, demonstrations, assistants, wizards, Internet support sites, interfaces) best fit which audience segments in which task domains with what product categories?
- How does the difference between initial learning, intermediate learning, advanced learning, and reference affect the choice of help systems and media of delivery?

(Continued)

TABLE A.1 (*continued*)

- How do ease-of-learning and ease-of-use differ in their effect on the choice of media and types of help/performance support?
- How do the choices of media and help/performance support change for complex tasks?
- What is the most effective ratio to use in media and types of help for specific kinds of cases?
- How should various media be ordered and organized for different types of help/performance support?
- What is the best mix of hard copy and online information?
- What are effective ways to organize information if you add graphics, animation, audio, and so on?
- What is the best way to integrate multimedia and improve the aesthetic and cognitive experience for users?
- How do you design hypertext for various purposes?
- What is best for a multiple audience? How can something be used broadly by different specific audiences?

Guidelines and Conventions

- What are effective guidelines for animation, sound, and multimedia?
- How are design conventions created or broken on the TV and Web?
- What are guidelines for designing Web documents?
- What are style standards for the Web?
- What are guidelines for distance learning?

Analyzing and Improving Usability

- What are differences between ease of learning and ease of use?
- How do users interact with data?
- How do people search?
- How do people choose?
- How do you instructionally support online decision making?
- What are problems in applying paper-based design criteria for online products?
- How can you help users with computer illiteracy?
- How can you make input devices effective?
- How can you make video on demand systems effective?
- How can you analyze the genre of TV/video? How can you merge that with the Web?

Support for Information Access, Retrieval, and Analysis

- How do you use new technologies to make users' legacy information and support their reasoning with it better?
- How do you help people search complex arrays so that they can make good decisions?
- How should graphics be designed for users trying to retrieve information from datacubes?
- How should data be designed to support complex retrieval and problem solving?
- What are ways to make information accessible?
- What is the optimum design for digital libraries?

Development

- What are strategies for making effective decisions about complex documents?
- What are strategies for effectively describing and presenting complex tasks and activities, especially when they require expertise in multiple task domains?
- How useful is redundancy in complex documents?
- What are strategies for determining what to revise first?
- What is the optimum set of information to include in a system library?

(Continued)

TABLE A.1 (*continued*)

International Communication

- Which international symbols work and which do not?
- What are key translation issues for technical communicators?

Production Tools

- How is cross-functional team work facilitated by group tools?
- When is it effective to move to single sourcing? When does it add or detract to use reusable objects?
- How do you evaluate new software for design activities?

Project Management

- How do you distribute support tasks and documentation jobs?
- How do you move documentation, online help, and media to one location?
- How can you avoid wasting time with Powerpoint presentations?
- When is it politically advantageous to save or spend money in a communication project?
- How does an organization know it's really doing better over time?
- How do consultants go into organizations and bring about change?
- How do you negotiate and navigate complex organizations?
- What are key issues and solutions related to intellectual property, archiving, version control, and archival control of copies?

Research and Practice

- How can you make research accessible to practitioners?
- How can you integrate current communication technology theory and practice in document design? How can you know when or where it is applicable to a particular context or situation? How much transfer is possible?

Literacies

- How can you give low or illiterate people health information they need?
- How can you motivate patients and clients to use health or social service documents? Does it help to use different forms and different language? Does it help to integrate more language elements?
- What are types of literacy and how can you build those into academic programs?

Global Conceptual Questions

- How do people deal with the complex, enormous Internet?
- What are strategies for facilitating experiences online?
- How can you apply principles of minimalism effectively to new contexts?
- What are challenges of complex systems?
- What kinds of support need to be built into complex systems?
- What will help or hinder people with products?
- What is the "value" of an interface? What makes a difference to this "value"?
- What are key branding issues and how can they be resolved?
- How do organizations capture knowledge and expertise? How do they know that they know? What is their rationale?
- What has to be done to make communication more accessible to include more groups? Do design choices make a difference?

About the Contributors

Lori Anschuetz is a principal of Tec-Ed, Inc., a 15-person firm specializing in usability research and information design. Located in the firm's Rochester (NY) office, she plans, implements, and manages projects for clients such as Sun Microsystems, Xerox, IEEE, and Cisco Systems. She has over 25 years of experience in corporate and consulting environments, including positions with ADP Network Services and Bendix Research Laboratories. A senior member of STC, Lori coauthored a chapter on audience analysis and document planning for the second edition of *Science and Technical Writing: A Manual of Style* (Routledge) edited by Philip Rubens. She has also coauthored papers for the Usability Professionals' Conference and IEEE International Professional Communication Conference, whose program committee she cochaired in 1996.

Stephen Bernhardt, Professor and Andrew B. Kirkpatrick Chair in Writing at the University of Delaware, is Past President of the Association of Teachers of Technical Writing (ATTW) and Council for Programs in Technical and Scientific Communication (CPTSC). Steve has also served on the Editorial Board of *Technical Communication Quarterly* and the *Journal of Business and Technical Communication,* and was Director of two National Workplace Literacy Demonstration Projects, funded by the U.S. Department of Education. Steve consults for global pharmaceutical companies on redesigning drug applications, developing training programs, creating intelligent systems, and creating communicative cultures to leverage organizational knowledge; he has also consulted and provided training to employees of IBM, Motorola, Hughes Aircraft, and other organizations. His research interests center on visual rhetoric, computers and writing, workplace training and development, and the teaching of scientific and technical writing.

Ann Blakeslee, Professor of English at Eastern Michigan University, has worked as a medical writer in a hospital public relations department, a technical writer in two industrial advertising agencies, and a technical communications consultant. She also has taught business, technical, and professional writing for 17 years at Eastern Michigan University, the University of Illinois–Urbana, Carnegie Mellon University, Villa Maria College, Gannon University, and Miami University of Ohio. At Eastern Michigan, she is Director of Writing Programs and Director of Writing Across the Curriculum. She also coordinates the Communications Outreach Program. In her scholarship, Ann has focused on genre theory, the rhetorical practices of scientists, and strategies for learning professional and academic genres through mentoring relationships and through classroom–workplace collaborations. She is author of *Interacting with Audiences: Social Influences on the Production of Scientific Writing* (Lawrence Erlbaum Associates) and Treasurer and Executive Committee Member of the Association of Teachers of Technical Writing (ATTW).

Russell Borland, retired Master Technical Writer, Microsoft Corporation, has 20 years of industry experience, mostly at Microsoft Corporation. After writing for Seafirst Bank in Seattle in 1979, Russell was first hired at Microsoft as a technical writer. In addition to writing manuals and product brochures, Russell supplied "information" for the MS-DOS 1.0 manual for the first IBM Personal Computer. In 1982, he became Manager of Technical Writing; in 1983, he was promoted to Manager of Technical Publications; and in 1984 he joined a team to design and develop Word for Windows version 1. In 1988, Russell transferred to Microsoft Press to write *Working with Word for Windows,* a book he has since revised several times under the title *Running Microsoft Word for Windows.* Borland was then promoted to Master Writer and authored or coauthored many books for Microsoft Press before retiring in 1997.

Deborah Bosley is Director of University Writing Programs and Associate Professor, University of North Carolina–Charlotte. She has taught for 19 years at UNCC, Millikin University, and Richland Community College, has consulted since 1985 for government, industry, and health organizations, and has served as Marketing Director for the Illinois Literary Publishers Association and Executive Director of Illinois Writers', Inc. Deborah was Member-at-Large for the Association of Teachers of Technical Writing (ATTW), Series Editor for the STC/ATTW Publication Series on Teaching Technical Communication, and Program Chair and Executive Committee Member for the Council for Programs in Technical and Scientific Communication (CPTSC). Deborah is editor of *Global Documentation: Case Studies in International Technical Communication* (Allyn and Bacon) and coeditor of *Teaching Strategies: Tips for Technical Communication Teachers* (STC) and *Technical Communication at Work* (Harcourt Brace). She has won the Nell Ann Picket Award for Best Article of the Year and the Distinguished Technical Communication Award for Outstanding Article.

Stanley Dicks, Assistant Professor at North Carolina State University, at one time guest lectured at Fairleigh Dickinson University and earned tenure and chaired the English Department at Wheeling Jesuit University, but he has spent most of his career, 17 years, in industry as a technical writer, proposals supervisor, and manager of documentation and training. He has worked for United Technologies, Burns and Roe, AT&T, and Bellcore. At NCSU, he teaches undergraduate and graduate courses in technical communication management and usability testing. His scholarly interests include integrated online support systems, usability testing, and user-centered design of technical documents, technical communication management principles and practices, and multimedia systems theory and design.

Brenton Faber is Assistant Professor at Clarkson University, where his scholarship focuses on organizational learning, organizational change, and professional communication. He is also studying corporate universities and firm-wide learning in for-profit and not-for-profit organizations. Brenton has worked for the Government of Ontario, has served as a volunteer executive director with Habitat for Humanity, and has been an independent consultant, specializing in issues of organizational change and company-wide learning. He is author of *Community Action and Organizational Change: Image, Narrative, Identity* (Southern Illinois University Press) and is working on a second book, tentatively entitled "Corporate Universities and Corporate Innovation: The Paradox of Organizational Learning."

Johndan Johnson-Eilola, Director of the Eastman Kodak Center for Excellence in Communication and Associate Professor at Clarkson University, has taught at Clarkson, the New Mexico Institute of Mining and Technology (as Director of Technical Communication), and Michigan Technological University (as a Ford Motor Company Doctoral Fellow). In addition to hypertext applications and theory, Johndan's scholarship has focused on finding intersections between practice and theory and examining how users learn to work within information saturated spaces. Johndan is author of *Nostalgic Angels: Rearticulating Hypertext Writing* (Ablex/Greenwood) and *Designing Effective Websites* (Houghton Mifflin) and coauthor of *Professional Writing Online* (Longman). He has won the Ellen Nold Award for Best Article, Nell Ann Picket Award for Best Article of the Year, and Hugh Burns Award for Best Dissertation in Computers and Composition Studies. He has also been keynote speaker at several rhetoric and technical communication conferences.

Barbara Mirel, Adjunct Professor at the University of Michigan and an industry consultant, has worked extensively in both academia and industry. She has 12 years of combined experience as Assistant Professor at Illinois Institute of Technology and Associate Professor at DePaul University. At DePaul, with Computer Science colleagues, she codeveloped and codirected the B.A. Program in Computing. As Senior Manager of Human Factors and a Member of the Technical Staff (MTS) at Lucent Technologies, Barbara focused on the design and usability of interactive data visualization programs and of new voice-over Internet telecommunications. As Cognitive Engineer for the National Center for Patient Safety, she led a field team investigating nationwide the usability and effectiveness of a patient bar code medication system and reported recommendations to the U.S. Under Secretary of Health. Barbara has published extensively in major journals and anthologies in the field. She received the SIGDOC Rigo award for lifetime achievement in 2000 and is an Associate Fellow of STC.

Anthony Paré, Chair and Associate Professor, Department of Educational Studies, McGill University, has taught for more than two decades and has conducted several ethnographic studies of social work writing. His scholarly interests have focused on workplace writing, school-to-work transitions, literacy education, and language and learning across the curriculum. Anthony coauthored *Worlds Apart: Acting and Writing in Academic and Workplace Contexts* (Lawrence Erlbaum Associates) and is coeditor (with Patrick Dias) of *Transitions: Writing in Academic and Workplace Settings* (forthcoming, Hampton). He was Director of McGill's Centre for the Study and Teaching of Writing, has also been active in the Canadian-based group of writers, *Inkshed,* for over a decade, serves on the Editorial Advisory Boards of *Textual Studies in Canada* and the Human Sciences Monograph Series of Laurentian University, and is a reviewer for the Social Sciences and Humanities Research Council of Canada.

Stephanie Rosenbaum is founder and president of Tec-Ed, Inc., a 15-person firm specializing in usability research and information design and headquartered in Ann Arbor, Michigan. Tec-Ed clients include AOL/Netscape, Rational Software, Intuit, Cisco Systems, and a wide variety of smaller firms. An STC Exemplar, Stephanie headed STC's Research Grants Committee for 5 years and has served on the STC Trends Committee. She was recently awarded a Millennium Medal by the IEEE Professional Communication Society. Her research background includes anthropology studies at Columbia University and experimental psychology research for the University of California at Berkley. Stephanie recently contributed a chapter to the Copenhagen Business School Press volume, *Software Design and Usability.* Her publications also include a chapter in John Carroll's book on *Minimalism Beyond the Nurnberg Funnel* and a variety of papers for conferences of ACM SIGCHI, ACM SIGDOC, and the Usability Professionals' Association.

Karen Schriver is President of KSA, Inc., a consulting firm that helps organizations improve the quality of print and electronic communications. Her clients have included Sony, Microsoft, United Chemicals of Belgium, Apple, IBM, ATT, Sprint, Mitsubishi, Sanyo, F. Hoffman-LaRoche (Switzerland), the Dutch Ministry of Education, and the Japan Management Association. Winner of six awards for her work and a popular speaker and keynote speaker, Karen specializes in information design and applying theory and research to solving everyday problems of communication. Karen is author of *Dynamics in Document Design: Creating Texts for Readers* (Wiley & Sons) and is now writing a book about the nature of expertise in information design. She was Codirector of Carnegie Mellon University's graduate program in professional communication and also held the Belle van Zuylen Professorship at the University of Utrecht in the Netherlands and a visiting position at the University of Washington. She is also an STC Associate Fellow and on the Editorial Board of *Technical Communication* and *Document Design* (a Dutch journal).

Rachel Spilka, Associate Professor at the University of Wisconsin–Milwaukee, has taught for two decades at UWM, Purdue University, the University of Maine–Orono (where she was Director of Technical Communication), Carnegie Mellon University, and Boston University. Rachel has also worked a combined 7 years in industry, first as a medical and technical writer, and more recently as a Technical Editor at MicroSym Corporation, a Senior Information Design Specialist at the American Institutes for Research, and a Communications Analyst at RAND Corporation. Rachel is editor of *Writing in the Workplace: New Research Perspectives* (Southern Illinois University Press), which won the NCTE Award for Excellence in Technical and Scientific Writing for Best Collection of Essays. She also won the same award for Best Article Reporting Formal Research. Rachel was Book Review Editor for *Technical Communication Quarterly* and Member-at-Large for the Association of Teachers of Technical Writing (ATTW), and is now Coordinator of the ATTW Research Committee.

Author Index

Note: An *f* or *t* immediately following a page number indicates a figure or table; an *n* immediately following a page number indicates a footnote on that page. Page numbers in italics indicate pages where full bibliographical information appears.

A

Abelson, R. P., 120, 120*n*, *129*
Adam, C., 41, 43, *54*, 61, 62, *71*, *72*
Agre, P., 170, 179, *186*
Alexander, C., 178, *186*
Ancell, N. S., 28, *39*
Angel, S., 178, *186*
Anson, C. M., 44, *54*
Applebee, A., 46, *54*
Ausubel, D., 126, *129*

B

Barchilon, M. G., 16, *25*
Bataille, R. R., 58, *71*
Bazerman, C., 58, *71*
Beck, K., 181, *186*
Berkenkotter, C., 43*n*, 48, *54*, 120, *129*, 177, *186*
Bernhardt, S., 101, *109*
Bever, H., 171, *186*
Blakeslee, A., 45, *54*
Blyler, N. R., 81, *90*
Borland, R., 96, 184, 189–195
Bosley, D., 19, *25*, 27, 28, 38, *39*
Bourdieu, P., 60, *71*
Brent, D., 60, *73*
Britton, W. E., 102, *109*
Brydon-Miller, M., 65, *72*
Burnett, R. E., 44, *54*

C

Caldas-Coulthard, C. R., 63, *71*
Carliner, S., 35–36, *39*, 106, *109*, 139, 140, *147*
Carroll, J. M., 89, *90*, 169, 186
Clark, R., 63, *71*
Cochran, C., 120, *129*
Cochran-Smith, M., 46, *54*
Coe, R., 64*n*, *71*
Cole, M., 61, *71*
Cooper, A., 171, *186*, 191, 193, *195*
Cooper, M. M., 58, *71*
Cope, B., 64*n*, *71*
Coplien, J., 178, *186*
Coulthard, M., 63, *71*

D

Davis, M. T., 99, 100, 106, 107, *109*
De Montigny, G., 66, *71*
Dias, P. X., 59, 59*n*, 61, 62, *71*
Dobrin, D. N., 102, *109*
Duin, A. H., 52, *54*

E–F

Ecker, P. S., 140, *148*
Faigley, L., 58, *71*
Fairclough, N., 63, *71*

Subject Index

Note: An *f* or *t* immediately following a page number indicates a figure or table; an *n* immediately following a page number indicates a footnote on that page. Page numbers in italics indicate pages where full bibliographical information appears.

A

Academia and industry
collaborations, 8–10, 12, 14, 15, 17, 20–25, 29–31, 44–54, 85–90
differences between, ix, 7, 8, 13–25, 28, 41, 43, 44, 53, 57, 81–85
cultural barriers, 7, 8, 10, 13, 14, 16, 28, 41–81, 85, 191
issues of funding, 83
issues of power, 20–21
issues of trust, 20, 81, 81*n*, 82–84
perceptions regarding,
assumptions about employment, 8, 14, 17, 23
collaboration, 14, 16–17
goals, ix, 7, 9, 18, 22, 24, 81
information, 8, 14–15, 23
language and discourse styles, 8, 14–16, 23, 44, 83
reward structures, ix, 14, 22–23, 83
time, ix, 18, 20–21
value of research, 85
workload, 18–20, 83
perspectives of, xi, 32–39, 82–85
philosophies, 21–23
prestige, 23
productive tension between, 89–90
structural barriers, 82–84
interactions
importance of, 13, 34

strategies for improving, xii
relationship, 2, 3, 9, 11, 27, 108
mutual benefits of, 13
fears and distrust of mutual influence, 81–86, 81*n*, 89–90
similarities, 7–9, 27–39, 43, 53
administration and management, 29, 31
collaboration, 29–30
education and training, 29, 31
foundations of expertise, 29–31
goals, 53
language, 53
need for research, 29
overlapping boundaries, 41–54
status, 29, 31–32
text and visual products, 29, 30
stereotyped perceptions of, 3, 30, 81, 82
strategies for improving the relationship between
active-practice, 81–90
contributing to each other's welfare, 9, 28
corporate–university hybrid, 138–147
identifying overlapping space, 9
making research accessible to industry, 36–39
preparing students to become reflective practitioners and practitioner researchers, 52
recognizing similarities, 27–32
relying on short-term collaborations, 14, 18, 24–25

213